Stochastic Partial Differential Equations with Additive Gaussian Noise

Analysis and Inference

T0321171

Stochastic Partial Differential Equations with Additive Gaussian Noise

Analysis and Inference

CIPRIAN A TUDOR

University of Lille, France

World Scientific

NEW JERSEY · LONDON · SINGAPORE · BEIJING · SHANGHAI · HONG KONG · TAIPEI · CHENNAI · TOKYO

Published by

World Scientific Publishing Co. Pte. Ltd.

5 Toh Tuck Link, Singapore 596224

USA office: 27 Warren Street, Suite 401-402, Hackensack, NJ 07601

UK office: 57 Shelton Street, Covent Garden, London WC2H 9HE

Library of Congress Control Number: 2022045674

British Library Cataloguing-in-Publication Data
A catalogue record for this book is available from the British Library.

**STOCHASTIC PARTIAL DIFFERENTIAL EQUATIONS WITH
ADDITIVE GAUSSIAN NOISE
Analysis and Inference**

ISBN 978-981-126-445-0 (hardcover)
ISBN 978-981-126-446-7 (ebook for institutions)
ISBN 978-981-126-447-4 (ebook for individuals)

For any available supplementary material, please visit
https://www.worldscientific.com/worldscibooks/10.1142/13089#t=suppl

Printed in Singapore

To Diana, my wife

Preface

The stochastic partial differential equations (in short, SPDEs) arise in many applications of the probability theory. Their study has a long history and many textbooks and monographs treat them in details. See, among many others, [Chow (2014)], [Da Prato and Zabczyk (1992)], [Liu and Röckner (2015)] or [Oksendal (2003)] for various methods to solve these equations and for a detailed study of the solutions to stochastic partial differential equations.

This monograph does not intend to describe how to solve the SPDEs in their very general form. We will focus on two particular (and probably the most known) equations: the stochastic heat equation and the stochastic wave equation. The random noise that drives these equations is always Gaussian and linear (or additive), i.e. the diffusive component of the equation does not depend on the solution itself. In this case, the proof of the existence of the solution does not require sophisticated techniques.

We more focus our attention on the relationship between the solutions to our SPDEs and the fractional Brownian motion (and related processes). The fractional Brownian motion (fBm) is a well-known stochastic process, widely studied in the recent and less recent years. It is a self-similar stochastic process with stationary increments. It generalizes the more famous Brownian motion but it is neither a semimartingale nor a Markov process. There exists a vast literature on fBm, see among others, [Mishura (2008)], [Nourdin (2012)], [Nualart (2006)]. This stochastic process, together with some of its generalizations and extension, is presented in our text but its presentation remains basic. Instead, we will describe in detail how the solutions to the stochastic heat and wave equations with additive Gaussian noise is connected with fractional Brownian motion or bifractional Brownian motion (an extension of the fBm) and we will use this connection in order to derive several distributional and trajectorial properties of these solutions. An important point of our analysis is the study of the asymptotic behavior of the p-variations of the solutions to the heat or wave equations driven by space-time Gaussian noise or by a Gaussian noise with a nontrivial correlation in space. The p-variations are then used to construct and analyze certain estimmators for the parameters that may appear in our stochastic models.

We start in Part 1 by a general brief description of the Gaussian processes and then we focus on fractional Brownian motion and some related processes. We also include a detailed analysis of some multiparameter (anisotropic and isotropic) Gaussian processes such as the fractional Brownian sheet, the space-time white noise or the white-colored Gaussian noise. In Part 2 we explore the link between the solutions to SPDEs and the multiparameter fractional Brownian motion. This link appears when we analyze the solution with respect to its time variable and as well with respect to its spatial variable. The last part of the monograph contains the study of the p-variations of the solutions to the stochastic heat and wave equations and the application of the p-variations to statistical inference. We describe the method to construct estimators for various parameters in our SPDEs (with focus on the estimation of the drift parameter) and we discuss the asymptotic behavior of these estimators.

The idea of the book (which is addressed to a public with a reasonable background in probability theory) is to keep it self-contained and to avoid the use of complex techniques (Itô integration, Wiener chaos, Malliavin calculus etc). We also chose to insist on the basic properties of the random noise (space-time white noise and white-colored Gaussian noise) and to detail the construction of the Wiener integration with respect to them. The intention is to make the proofs complete and pretty detailed.

Ciprian A. Tudor
Paris, April 2022

Contents

Preface   vii

Gaussian processes and sheets   1

1. Gaussian processes   3

    1.1   Generalities on stochastic processes   3
    1.2   Gaussian random variables and vectors   8
    1.3   Gaussian processes: basic properties   10

2. Fractional and Bifractional Brownian motion   15

    2.1   Fractional Brownian motion   15
    2.2   Bifractional Brownian motion   25

3. Multiparameter Gaussian processes   33

    3.1   The anisotropic fractional Brownian sheet   33
    3.2   The space-time white noise   37
          3.2.1   The space-time white noise and the anisotropic
                  Brownian sheet   39
    3.3   Isotropic fractional Brownian sheet   41
    3.4   The white-colored Gaussian noise   43
          3.4.1   The white-colored Gaussian noise and the
                  anisotropic fractional Brownian sheet   45
          3.4.2   Other spatial covariance kernels   47

4.   Isonormal processes and Wiener integral   49

   4.1   Definition and general properties   49
   4.2   The Brownian motion as an isonormal process   53
   4.3   The fractional Brownian motion as an isonormal process .   56
   4.4   The space-time white noise as an isonormal process   61
   4.5   The white-colored noise as an isonormal process   65

**Stochastic heat and wave equations with additive
Gaussian noise**   **69**

5.   The stochastic heat equation with space-time white noise   71

   5.1   The standard stochastic heat equation with space-time
         white noise   72
         5.1.1   General properties of the solution   72
         5.1.2   Relation with the weak solution   76
         5.1.3   Analysis of the solution in time   77
         5.1.4   Behavior in space   84
   5.2   The fractional stochastic heat equation with space-time
         white noise   92
         5.2.1   General properties   92
         5.2.2   Behavior in time   95
         5.2.3   Behavior in space   98

6.   The stochastic heat equation with correlated noise in space   103

   6.1   The standard stochastic heat equation with white-colored
         noise 103
         6.1.1   General properties 103
         6.1.2   Behavior in time 106
         6.1.3   Behavior in space 110
   6.2   The fractional heat equation with white-colored noise . . . 114
         6.2.1   General properties 114
         6.2.2   Behavior in time 115
         6.2.3   Behavior in space 118

7.   The stochastic wave equation with space-time white noise   123

   7.1   General properties 123
         7.1.1   Relation with the weak solution 130

7.2 Behavior of the solution in time 134
7.3 Behavior of the solution in space 137

Power variation and statistical inference for solutions to SPDEs 143

8. Variations of the solution to the stochastic heat equation 145

8.1 Exact and renormalized variations of the perturbed
   fractional Brownian motion 145
8.2 Variations of the isotropic fractional Brownian motion . . 152
8.3 Variations of the solution to the heat equation with
   space-time white noise 154
   8.3.1 Temporal variations 155
   8.3.2 Spatial variations 157
8.4 Variations of the solution to the heat equation with
   white-colored noise 159

9. Parameter estimation for the stochastic heat equation via
   power variations 161

9.1 The parametrized heat equation with space-time
   white noise 162
   9.1.1 Properties of the solution 163
   9.1.2 Estimators of the drift parameter 166
9.2 The parametrized heat equation with correlated spatial
   noise 168
   9.2.1 Properties of the solution 169
   9.2.2 Estimators of the drift parameter 172

10. Power variations and inference for stochastic the wave equation 175

10.1 Power variation of the solution 175
    10.1.1 Temporal variation 175
    10.1.2 Spatial variation 179
10.2 The parametrized wave equation with space-time white
    noise 182
    10.2.1 Properties of the solution 182
    10.2.2 Estimation of the drift parameter 185

Bibliography 189

PART 1
Gaussian processes and sheets

Chapter 1

Gaussian processes

In this part, we introduce the basic concepts on random variables and stochastic processes needed in the next chapters.

1.1 Generalities on stochastic processes

Let $(\Omega, \mathcal{F}, P)$ be a probability space and let T be a non-empty set. A random variable X is a $\mathcal{F}$-measurable function $X : \Omega \to \mathbb{R}$.

A (real-valued) stochastic process $X = (X_t, t \in T)$ (also denoted by $X = (X_t)_{t \in T}$) is a collection of random variables on $(\Omega, \mathcal{F}, P)$, i.e. for every $t \in T$, $X_t : \Omega \to \mathbb{R}$ is a $\mathcal{F}$-measurable mapping. Often we will consider $T = [0, \infty)$, $T = \mathbb{R}$ or $T = [a, b]$ with $a < b$. We use the following terminology:

- If T is finite, then X is a random vector.
- If $T \subset \mathbb{Z}$, then Z is a discrete stochastic process.
- It $T \subset \mathbb{R}^d$, with $d \geq 2$, then $X = (X_t, t \in T)$ is a random field.

A filtration $(\mathcal{F}_t, t \in T)$ is a family of sub-sigma-algebras of $\mathcal{F}$ such that $\mathcal{F}_s \subset \mathcal{F}_t$ if $s \leq t$. We say that the stochastic process $X = (X_t, t \in T)$ is adapted to the filtration $(\mathcal{F}_t, t \in T)$ (or simply, X_t is $\mathcal{F}_t$-adapted) if the random variable X_t is measurable with respect to the sigma-algebra $\mathcal{F}_t$ for every $t \in T$.

Recall that the probability law (or the probability distribution) of the random vector $X = (X_1, \ldots, X_d)$ is usually denoted by P_X (or $P \circ X^{-1}$) and it is defined by

$$P_X(A) = (P \circ X^{-1})(A) = P(X \in A) = P(X_1 \in A_1, \ldots, X_d \in A_d)$$

for any $A = A_1 \times A_2 \times \ldots \times A_d \in \mathcal{B}(\mathbb{R}^d)$. The application $P_X : \mathcal{B}(\mathbb{R}^d) \rightarrow [0,1]$ is a probability measure on $(\mathbb{R}^d, \mathcal{B}(\mathbb{R}^d))$. We denote by $\mathcal{B}(U)$ the Borel sets of the topological space U.

Definition 1.1. The finite dimensional distributions of the process $X = (X_t, t \in T)$ represent the totality of the probability distributions of the random vectors

$$\left(X_{t_1}, \ldots, X_{t_p}, t_1, \ldots, t_p \in T, p \geq 1 \right).$$

In the sequel we will say that *two stochastic processes have the same law* if they have the same finite-dimensional distributions.

A trajectory of the real-valued stochastic process $(X_t, t \in T)$ is a mapping $t \rightarrow X_t(\omega)$ from T onto $\mathbb{R}$ with $\omega \in \Omega$. It is always of interest to analyze the regularity of the trajectories for a given stochastic process. We say that a process X is continuous if all its trajectories (also called sample paths) are continuous.

We introduce the concepts of *modification* and *indistinguishability* for stochastic processes.

Definition 1.2. Two stochastic processes $X = (X_t, t \in T)$ and $Y = (Y_t, t \in T)$ are called modifications (or versions) one of the other if for every $t \in T$,

$$P(X_t = Y_t) = 1.$$

We say that the processes X and Y are indistinguishable if

$$P(X_t = Y_t, \ \forall t \in T) = 1.$$

Let us make few comments concerning these two notions:

- If X and Y are indistinguishable, then they are modifications one of the other.
- The converse direction is not true. Take $T = [0,1]$ and $(\Omega, \mathcal{F}, P) = ([0,1], \mathcal{B}([0,1]), \lambda)$ with λ the Lebesgue measure. Define

$$X_t(\omega) = \begin{cases} 0, & \text{if } t \neq \omega \\ 1 & \text{if } t = \omega. \end{cases}$$

Then X, Y are modifications but are not indistinguishable. Indeed, for every $t \in T = [0,1]$,

$$P(w \in \Omega, X_t(\omega) \neq Y_t(\omega)) = \lambda(\{t\}) = 0$$

while

$$P\left(\omega \in \Omega, X_t(\omega) = Y_t(\omega) \text{ for every } t \in T\right)$$
$$= \lambda\left(\omega \in [0,1], \omega \neq t \text{ for every } t \in [0,1]\right)$$
$$= \lambda(\emptyset) = 0.$$

- If X and Y are modifications and their paths are (right)-continuous then they are indistinguishable.
- If two stochastic processes are modifications one of the other, then they have the same finite-dimensional distributions.
- The inverse direction is not true. Take $X_t = Z$ and $Y_t = -Z$ or every $t \geq 0$ with Z a random variable such that Z and $-Z$ have the same law (for example, if Z is a standard normal random variable).

The next result is many times used to obtain the continuity of sample paths of stochastic processes (we refer to it as to the *Kolmogorov continuity criterion*).

Theorem 1.1. *Consider a stochastic process $(X_t, t \in T)$ where $T \subset \mathbb{R}$ is a compact set. Suppose that there exist constant $p, C > 0$ and $\beta > 1$ such that for every $s, t \in T$*

$$\mathbf{E}\left|X_t - X_s\right|^p \leq C|t - s|^\beta.$$

Then X has a continuous modification $\tilde{X}$. Moreover for every $0 < \gamma < \frac{\beta-1}{p}$

$$\mathbf{E}\left(\sup_{s,t \in T; s \neq t} \frac{|\tilde{X}_t - \tilde{X}_s|}{|t - s|^\gamma}\right)^p < \infty.$$

In particular X admits a modification which is Hölder continuous of any order $\alpha \in (0, \frac{\beta-1}{p})$.

Let us now define several particular classes of stochastic processes.

Definition 1.3. Let $T = \mathbb{R}$ or $T = \mathbb{R}_+$. A stochastic process $(X_t, t \in T)$ is called (strictly) stationary if for any $t_1, \ldots, t_n \in T$ and for any $k \in T$, the joint distribution of the vector

$$(X_{t_1+k}, \ldots, X_{t_n+k})$$

coincides with the joint distribution of the random vector

$$(X_{t_1}, \ldots, X_{t_n}).$$

Similarly, if $(X_k, k \in \mathbb{Z})$ is a discrete stochastic process, it is (strictly) stationary if for every $k_1, \ldots, k_n, k \in \mathbb{Z}$, the random vectors

$$(X_{k_1+k}, \ldots, X_{k_n+k}) \text{ and } (X_{k_1}, \ldots, X_{k_n})$$

have the same finite dimensional distributions.

That means that the statistical properties of the process X do not change over time.

A stochastic process $(X_t, t \geq 0)$ is stochastically continuous at t if for every $\varepsilon > 0$,

$$\lim_{h \to 0} P\left(|X_{t+h} - X_t| > \varepsilon\right) = 0.$$

The process X is said to be trivial if $P \circ X_t^{-1}$ is a delta measure for every $t \geq 0$.

Definition 1.4. A stochastic process $X = (X_t, t \geq 0)$ is self-similar if for every $a > 0$ there exists $b > 0$ such that

$$(X_{at}, t \geq 0) \equiv^{(d)} (bX_t, t \geq 0).$$

The following facts hold true:

- if $(X_t, t \geq 0)$ is stochastically continuous at $t = 0$ and non-trivial, then there exists a unique index $H \geq 0$ such that $b = a^H$.
- moreover, $X_0 = 0$ almost surely if and only if $H > 0$.
- we say that the process X is H-self-similar.

Remark 1.1.

- We can also define the self-similar processes as follows (this definition can be found more often in the literature): a process $(X_t, t \geq 0)$ is self-similar is there exists an index $H > 0$ such that for every $c > 0$ the processes

$$(X_{ct}, t \geq 0) \text{ and } (c^H X_t, t \geq 0)$$

  have the same finite dimensional distributions. It can be shown that the exponent H is unique if the process X is stochastically continuous at $t = 0$ and non-trivial.
- We can define the concept of self-similarity if the time interval $\mathbb{R}_+$ is replaced by $\mathbb{R}$. That is, in the sequel we will say that a process $X = (X_t, t \in \mathbb{R})$ is H-self-similar with $H > 0$ if for every $c > 0$, $(X_{ct}, t \in \mathbb{R})$ and $(c^H X_t, t \in \mathbb{R})$ have the same law.
- A detailed presentation of the notion of self-similarity can be found in the monograph [Embrechts and Maejima (2002)].

Remark 1.2.

- There exists a different notion of stationarity called weak stationarity (of stationarity in the weak sense) which requires the first moment and the covariance function to not vary in time. That is, a process $(X_t, t \in T)$ $(T = \mathbb{R}_+, \mathbb{R}$ or $\mathbb{Z})$ is stationary in the weak sense if $\mathbf{E}X_{t+h} = \mathbf{E}X_t$ for every $t, h \in T$ and

$$Cov(X_t, X_s) = Cov(X_{t-s}, X_0) := Q(t - s) \text{ for every } s, t \in T,$$
(1.1)

  so the covariance of the process $Cov(X_t, X_s)$ is a function of $t - s$ for every $s \leq t$. If $T = \mathbb{R}$ or $T = \mathbb{Z}$, then $Q(-t) = Q(t)$ for every $t \in \mathbb{R}$ so Q is an even function.

- If the process is centered and Gaussian, the two notions (strict stationarity and weak stationarity) coincide, see Proposition 1.2 below.

- If $(X_t, t \in \mathbb{R})$ is a centered stationary stochastic process, then

$$\mathbf{E}(X_t - X_s)^2 = 2(Q(0) - Q(t - s)), \quad s, t \in \mathbb{R}$$
(1.2)

  with Q from (1.1). Indeed,

$$\mathbf{E}(X_t - X_s)^2 = \mathbf{E}X_t^2 + \mathbf{E}X_s^2 - 2\mathbf{E}X_tX_s$$
$$= 2Q(0) - 2Q(t - s).$$

- The covariance function of a stationary process attains its maximum at zero. This is true because by (1.2),

$$Q(0) - Q(t) \geq 0 \text{ for every } t \in \mathbb{R}.$$

A trivial example of stationary discrete process is the following.

Example 1.1.

(1) Let $(X_k, k \in \mathbb{Z})$ be a sequence of independent indentically distributed random variables with mean zero and variance $\sigma^2 < \infty$. Then the sequence $(X_k, k \in \mathbb{Z})$ is strongly stationary.

(2) Let A, B be two uncorrelated centered random variables with finite variance. For $w \in [0, 2\pi]$ define the process $X = (X_t, t \in \mathbb{R})$ by

$$X_t = A\cos(wt) + B\sin(wt), \quad t \in \mathbb{R}.$$

Then X is weakly stationary (in general it is not strongly stationary).

Another important class of stochastic processes is the class of processes with stationary increments.

Definition 1.5. Let $T = \mathbb{R}_+$ or $T = \mathbb{R}$. A process $X = (X_t, t \in T)$ has stationary increments if for every $h \in T$, the stochastic process

$$(X_{t+h} - X_h, t \in T)$$

has the same finite-dimensional distributions as the process X.

That means that the law of the increment process $(X_{t+h} - X_h, t \in T)$ does not depend on h.

The covariance of a self-similar process with stationary increments is always the same.

Theorem 1.2. *Let $X = (X_t, t \in \mathbb{R})$ be a (non-trivial) H-self-similar process with stationary increments and assume $\mathbf{E}X_1^2 < \infty$. Then for every $s, t \in \mathbb{R}$,*

$$\mathbf{E}X_t X_s = \frac{1}{2} \left(|t|^{2H} + |s|^{2H} - |t-s|^{2H} \right) \mathbf{E}X_1^2.$$

Proof. Take $s, t \in \mathbb{R}$ arbitrary chosen and recall that $X_0 = 0$ almost surely. Then by the stationarity of the increments

$$\mathbf{E}X_t X_s = \frac{1}{2} \mathbf{E} \left(X_t^2 + X_s^2 - |X_t - X_s|^2 \right)$$

$$= \frac{1}{2} \mathbf{E} \left(X_t^2 + X_s^2 - |X_{t-s}|^2 \right)$$

and by self-similarity

$$\mathbf{E}X_t X_s = \frac{1}{2} \mathbf{E} \left(|t|^{2H} \mathbf{E}X_1^2 + |s|^{2H} \mathbf{E}X_1^2 - |t-s|^{2H} \mathbf{E}X_1^2 \right)$$

$$= \frac{1}{2} \left(|t|^{2H} + |s|^{2H} - |t-s|^{2H} \right) \mathbf{E}X_1^2.$$

$\square$

Theorem 1.2 says that there exists a unique centered Gaussian self-similar process X with stationary increments and $\mathbf{E}X_1^2 = 1$. This is the fractional Brownian motion presented in Chapter 2.

1.2　Gaussian random variables and vectors

In the probability theory, the Gaussian random variable is clearly the most commonly used and of most importance. A random variable $X : (\Omega, \mathcal{F}, P) \to \mathbb{R}$ follows the Gaussian (or normal) distribution $N(m, \sigma^2)$ (we write $X \sim N(m, \sigma^2)$) if it admits a density given by

$$f(x) = \frac{1}{\sqrt{2\pi}\sigma} e^{-\frac{(x-m)^2}{2\sigma^2}}, \quad x \in \mathbb{R},$$

i.e. for every $a, b \in \mathbb{R}, a < b$

$$P\left(a \leq X \leq b\right) = \int_a^b f(x)dx = \frac{1}{\sqrt{2\pi}\sigma} \int_a^b e^{-\frac{(x-m)^2}{2\sigma^2}} dx.$$

If $X \sim N(0,1)$, we say that X is a standard Gaussian random variable.

We recall some basic facts on Gaussian random variables. Assume $X \sim N(m, \sigma^2)$.

- $\mathbf{E}X = m$ and $Var(X) = \sigma^2$.
- the characteristic function of X is given by

$$\varphi_X(t) = \mathbf{E}\left(e^{itX}\right) = e^{itm - \frac{\sigma^2 t^2}{2}}, \qquad t \in \mathbb{R}.$$

- if Y is another random variable such that $Y \sim N(m_1, \sigma_1^2)$ and Y is independent by $X \sim N(m, \sigma^2)$, then

$$X + Y \sim N(m + m_1, \sigma^2 + \sigma_1^2).$$

- If $m = 0$, then X has the following moments, for $n \geq 1$,

$$\mathbf{E}X^{2n} = \frac{(2n)!}{n! 2^n} \sigma^{2n} \text{ and } \mathbf{E}X^{2n+1} = 0.$$

Let us denote by $\langle \cdot, \cdot \rangle$ the Euclidean scalar product in $\mathbb{R}^d$. We introduce the Gaussian random vectors.

Definition 1.6. A random vector $X = (X_1, \ldots, X_d) : \Omega \to \mathbb{R}^d$ is Gaussian if for every $\lambda_1, \ldots, \lambda_d \in \mathbb{R}$, the random variable

$$\langle X, \lambda \rangle = \lambda_1 X_1 + \ldots \lambda_d X_d$$

is normally distributed.

The expectation of the vector X is the vector in $\mathbb{R}^d$

$$\mathbf{E}X = (\mathbf{E}X_1, \ldots, \mathbf{E}X_d)$$

while its covariance matrix is defined by $C = (C_{i,j})_{1 \leq i,j \leq d}$ with

$$C_{i,j} = Cov(X_i, X_j) \text{ for every } 1 \leq i, j \leq d.$$

We also recall some useful facts concerning the Gaussian vectors. Let $X = (X_1, \ldots, X_d) : \Omega \to \mathbb{R}^d$ be a Gaussian vector.

- For every $i = 1, \ldots, d$, the random variable X_i is Gaussian.

- If X_i is Gaussian for every $i = 1, \ldots, d$ and $X_i, 1 \leq i \leq d$ are (mutually) independent, then the vector $X = (X_1, \ldots, X_d)$ is Gaussian. The claim is not true without the assumption of independence.
- The law of the random vector X is completely determined by its expectation $\mathbf{E}X$ and its covariance matrix C. For instance, its characteristic function

$$\varphi_X(t) = \mathbf{E}\left(e^{i\langle t, X \rangle}\right), \quad t \in \mathbb{R}^d$$

can be written as, for $t = (t_1, \ldots, t_d)$,

$$\varphi_X(t) = e^{i\langle t, \mathbf{E}X \rangle - \frac{1}{2} t^T C t} = e^{i \sum_{j=1}^d x_j \mathbf{E}X_j - \frac{1}{2} \sum_{j,k=1}^d t_j t_k C_{j,k}(X)},$$

where "T" denote the transpose.
- The density of the random vector X is, if $\det(C) \neq 0$,

$$f_X(x) = \frac{1}{\sqrt{(2\pi)^d \det C}} e^{-\frac{1}{2}\langle (x-m)^T C^{-1}(x-m) \rangle}$$

for every $x \in \mathbb{R}^d$. In particular, if C is the $d \times d$ identity matrix I_d, then for every $x = (x_1, \ldots, x_d)$,

$$f_X(x) = \frac{1}{\sqrt{(2\pi)^d}} e^{-\frac{1}{2}(x_1^2 + \ldots + x_d^2)}.$$

- two components X_i and X_j $(1 \leq i, j \leq d, i \neq j)$ of the vector X are independent if and only if they are non-correlated, i.e.

$$Cov(X_i, X_j) = 0 \text{ if } i \neq j.$$

Notice that this property only holds for components of a Gaussian vector. For instance, if X, Y are Gaussian random variable such that $Cov(X, Y) = 0$, the random variables X and Y are not necessarily independent.

1.3 Gaussian processes: basic properties

Let T be a non-empty set. We start with the definition of a Gaussian process.

Definition 1.7. The stochastic process $X = (X_t, t \in T)$ is Gaussian if for every $\lambda_1, \ldots, \lambda_n \in \mathbb{R}$ and for every $t_1, \ldots, t_n \in T$, the random variable

$$\lambda_1 X_{t_1} + \ldots \lambda_n X_{t_n}$$

is Gaussian.

Otherwise, for every $t_1, \ldots, t_N \in T$, the random vector $(X_{t_1}, \ldots, X_{t_n})$ is Gaussian. We will also sometimes use the denomination Gaussian family if the process X is indexed by a non-empty set I which is not included in $\mathbb{R}^d$ (it may be an Hilbert space or a space of functions etc).

The expectation of a process $X = (X_t, t \in T)$ is the function

$$m_X : T \to \mathbb{R}, \quad m_X(t) = \mathbf{E}X_t$$

while its covariance function is the mapping $C_X : T^2 \to \mathbb{R}$ given by

$$C_X(t, s) = Cov(X_t, X_s) = \mathbf{E}(X_t X_s) - (\mathbf{E}X_t)(\mathbf{E}X_s)$$

for every $s, t \in T$.

Definition 1.8. A function $C : T^2 \to \mathbb{R}$ is said to be positive definite (or of positive type) if for every $N \geq 1$ and for every $\lambda_1, \ldots, \lambda_N \in \mathbb{R}, t_1, \ldots, t_N \in T$, one has

$$\sum_{i,j=1}^{N} \lambda_i \lambda_j C(t_i, t_j) \geq 0. \tag{1.3}$$

Proposition 1.1. *If C is the covariance function of a stochastic process $X = (X_t, t \in T)$ then C is symmetric and positive definite.*

Proof. Clearly for every $s, t \in T$,

$$C(t, s) = Cov(X_t, X_s) = Cov(X_s, X_t) = C(s, t)$$

so C is symmetric. Take $\lambda_1, \ldots, \lambda_N \in \mathbb{R}, t_1, \ldots, t_N \in T$ arbitrary. Then

$$\sum_{i,j=1}^{N} \lambda_i \lambda_j C(t_i, t_j) = \sum_{i,j=1}^{N} \lambda_i \lambda_j Cov(X_{t_i}, X_{t_j})$$

$$= Var\left(\sum_{i=1}^{N} \lambda_i X_{t_i}\right) \geq 0$$

so C is of positive type. □

Remark 1.3.

- If $C : T^2 \to \mathbb{R}$ is symmetric and positive definite, then

$$|C(t, s)|^2 \leq C(t, t)C(s, s) \text{ for every } s, t \in T.$$

This is the Cauchy-Schwarz inequality. In particular, $C(t, t) \geq 0$ for every $t \in T$.

- Given two covariance functions (symmetric and positive definite) $C_1, C_2 : T^2 \to \mathbb{R}$, then the following are also covariance functions:

$$C_1 + C_2, \quad C_1 C_2, \quad e^{C_1} - 1.$$

A proof can be found in e.g. [Nourdin (2012)].

Given a covariance function (symmetric and positive definite), we can always construct a Gaussian process with such a covariance. The next result is due to Kolmogorov (see e.g. [Kolmogorov (1950)], [Dudley (2003)]).

Theorem 1.3. *Let $C : T^2 \to \mathbb{R}$ be symmetric and of positive type. Then there exists a centered Gaussian process $(X_t, t \in T)$ such that for every $s, t \in T$,*

$$\mathbf{E} X_t X_s = C(t, s).$$

The expectation and the covariance function determine entirely the law of a Gaussian process.

Theorem 1.4. *Two Gaussian processes with the same expectation and the same covariance function have the same finite-dimensional distributions.*

For stationary Gaussian processes, we have the following characterization.

Proposition 1.2. *Let $(X_t, t \geq 0)$ be a Gaussian process. Then the following are equivalent.*

(1) X is is strictly stationary.
(2) X is weakly stationary.
(3) there exists a function $C : [0, \infty)^2 \to \mathbb{R}$ such that the covariance function of the process X, $R(s, t) = Cov(X_t, X_s)$, satisfies

$$R(t, s) = C(t - s) \quad \text{for every } s, t \geq 0.$$

Proof. The equivalence between the points 1. and 2. follows from the fact that the covariance function determines the law of a Gaussian process (Theorem 1.4). The equivalence between 2. and 3. is due to the following equality

$$\mathbf{E} X_{t+h} X_{s+h} = \mathbf{E} X_t X_s$$

for every $s, t \geq 0$ and for every $h > 0$. $\qquad\qquad\qquad\qquad\qquad\square$

Example 1.2. A trivial example of a Gaussian process is $(X_t, t \geq 0)$ defined by

$$X_t = tZ, \quad t \geq 0$$

where $Z \sim N(0,1)$. This is a centered Gaussian process with covariance function

$$\mathbf{E}X_t X_s = ts\mathbf{E}Z^2 = ts$$

for every $s, t \geq 0$. Many other Gaussian processes will be treated in the next chapters.

Example 1.3. Let $(\xi_k, k \in \mathbb{Z})$ be a sequence of i.i.d $N(0,1)$ distributed random variables. Then $(\xi_k, k \in \mathbb{Z})$ is a discrete stationary Gaussian process.

Chapter 2

Fractional and Bifractional Brownian motion

We will discuss in details some Gaussian processes which are linked with the solutions to the stochastic partial differential equations presented in the next chapters. These processes will be generically called "fractional processes" because they are all somehow related to the fractional Brownian motion.

2.1 Fractional Brownian motion

Let us start with the following result which justifies the existence of the (one-parameter or multiparameter) fractional Brownian motion. Recall that $\langle \cdot, \cdot \rangle$ and $\| \cdot \|$ denotes the Euclidean scalar product and the Euclidean norm in $\mathbb{R}^d$, respectively.

Proposition 2.1. *Let $H \in (0, 1]$. Consider the function $C : T^2 = (\mathbb{R}^d)^2 \to \mathbb{R}$ given by*

$$C(t, s) = \frac{1}{2} \left(\|t\|^{2H} + \|s\|^{2H} - \|t - s\|^{2H} \right), s, t \in \mathbb{R}^d. \qquad (2.1)$$

Then C is symmetric and of positive type.

Proof. Assume first $H \in (0, 1)$. Clearly C is symmetric. We use the formula, for $\lambda > 0$, $0 < K < 1$,

$$\lambda^K = \frac{K}{\Gamma(1 - K)} \int_0^\infty (1 - e^{-\lambda x}) x^{-1-K} dx. \qquad (2.2)$$

Indeed,

$$\int_0^\infty (1 - e^{-\lambda x}) x^{-1-K} dx$$

$$= \frac{1}{\lambda} \int_0^\infty ds \left(\int_0^x dy e^{-\lambda y} \right) x^{-1-k}$$

$$= \frac{1}{\lambda} \int_0^\infty dy e^{-\lambda y} \int_y^\infty dx x^{-1-K} = \frac{1}{\lambda K} \int_0^\infty dy e^{-\lambda y} y^{-K}$$

$$= \frac{\Gamma(1-K)}{K} \lambda^K.$$

By (2.2), we have for every $\lambda \in \mathbb{R}$,

$$|\lambda|^{2H} = \frac{H}{\Gamma(1-H)} \int_0^\infty \left(1 - e^{-\lambda^2 x} \right) x^{-1-H} dx$$

$$= c(H) \int_0^\infty (1 - e^{-u^2 |\lambda|^2}) u^{-1-2H} du \qquad (2.3)$$

with $c(H) = 2\frac{H}{\Gamma(1-H)}$. We apply (2.3) for $\lambda = \|t\|, \lambda = \|s\|$ and $\lambda = \|t-s\|$ with $t, s \in T = \mathbb{R}^d$. Thus, for $t, s \in \mathbb{R}^d$ arbitrary,

$$\|t\|^{2H} + \|s\|^{2H} - \|t-s\|^{2H}$$

$$= c(H) \left[\int_0^\infty \left(1 - e^{-u^2 \|t\|^2} \right) u^{-1-2H} du + \int_0^\infty \left(1 - e^{-u^2 \|s\|^2} \right) u^{-1-2H} du \right.$$

$$\left. - \int_0^\infty \left(1 - e^{-u^2 \|t-s\|^2} \right) u^{-1-2H} du \right].$$

By using the identity

$$\left(1 - e^{-u^2 \|t\|^2} \right) + \left(1 - e^{-u^2 \|s\|^2} \right) - \left(1 - e^{-u^2 \|t-s\|^2} \right)$$

$$= \left(1 - e^{-u^2 \|t\|^2} \right) \left(1 - e^{-u^2 \|s\|^2} \right) + e^{-u^2 (\|t\|^2 + \|s\|^2)} \left(e^{2u^2 \langle t,s \rangle} - 1 \right)$$

we obtain

$$\|t\|^{2H} + \|s\|^{2H} - \|t-s\|^{2H}$$

$$= c(H) \left[\left(1 - e^{-u^2 \|t\|^2} \right) \left(1 - e^{-u^2 \|s\|^2} \right) u^{-1-2H} du \right.$$

$$\left. + \int_0^\infty e^{-u^2 (\|t\|^2 + \|s\|^2)} \left(e^{2u^2 \langle t,s \rangle} - 1 \right) u^{-1-2H} du \right]. \qquad (2.4)$$

We have, for every $s, t \in \mathbb{R}$,

$$e^{2\langle t,s \rangle u^2} - 1 = \sum_{n \geq 1} \frac{2^n \langle t, s \rangle^n}{n!} u^{2n}$$

and then, by (2.4),

$$\|t\|^{2H} + \|s\|^{2H} - \|t-s\|^{2H}$$

$$= c(H)^{-1} \left[\int_0^\infty \left(1 - e^{-u^2\|t\|^2}\right) \left(1 - e^{-u^2\|s\|^2}\right) u^{-1-2H} du \right.$$

$$\left. + \sum_{n\geq 1} \frac{2^n \langle t,s \rangle^n}{n!} \int_0^\infty u^{2n-2H-1} e^{-u^2(\|t\|^2+\|s\|^2)} du \right] \qquad (2.5)$$

Now, take $N \geq 1$, $\lambda_1,\ldots,\lambda_N \in \mathbb{R}$ and $t_1,\ldots,t_N \in \mathbb{R}^d$. We have from (2.5)

$$c(H) \sum_{i,j=1}^N \lambda_i \lambda_j \left(\|t_i\|^{2H} + \|t_j\|^{2H} - \|t_i - t_j\|^{2H} \| \right)$$

$$= \sum_{i,j=1}^N \lambda_i \lambda_j \int_0^\infty \left(1 - e^{-u^2\|t_i\|^2}\right) \left(1 - e^{-u^2\|t_j\|^2}\right) u^{-1-2H} du$$

$$+ \sum_{i,j=1}^N \lambda_i \lambda_j \sum_{n\geq 1} \frac{2^n \langle t_i, t_j \rangle^n}{n!} \int_0^\infty u^{2n-2H-1} e^{-u^2(\|t_i\|^2+\|t_j\|^2)} du$$

$$:= A + B.$$

We notice that

$$A = \int_0^\infty \left(\sum_{i=1}^N \lambda_i \left(1 - e^{-u^2\|t_i\|^2}\right) \right)^2 u^{-1-2H} \geq 0$$

and

$$B = \sum_{n\geq 1} \frac{2^n}{n!} \int_0^\infty du\, u^{2n-H-1} \sum_{i,j=1}^N \lambda_i \lambda_j \langle t_i, t_j \rangle^n e^{-u^2(\|t_i\|^2+\|t_j\|^2)}$$

$$= \sum_{n\geq 1} \frac{2^n}{n!} \int_0^\infty du\, u^{2n-H-1} \sum_{i,j=1}^N \lambda_i \lambda_j \langle \left(e^{-u^2\|t_i\|^2}\right)^{\frac{1}{n}} t_i, \left(e^{-u^2\|t_j\|^2}\right)^{\frac{1}{n}} t_j \rangle^n$$

$$= \sum_{n\geq 1} \frac{2^n}{n!} \int_0^\infty du\, u^{2n-H-1} \left\| \sum_{i=1}^N \lambda_i \left(e^{-u^2\|t_i\|^2}\right)^{\frac{1}{n}} t_i \right\|^{2n} \geq 0.$$

So the result is obtained for $H \in (0,1)$.

Let $H = 1$. Then

$$C(t,s) = \frac{1}{2}(\|t\|^2 + \|s\|^2 - \|t-s\|^2) = \langle t,s \rangle.$$

For $\lambda_1, \ldots, \lambda_N \in \mathbb{R}$ and $t_1, \ldots, t_N \in \mathbb{R}$,

$$\sum_{i,j=1}^{N} \lambda_i \lambda_j C(t_i, t_j) = \sum_{i,j=1}^{N} \lambda_i \lambda_j \langle t_i, t_j \rangle$$

$$= \left\| \sum_{j=1}^{N} \lambda_j t_j \right\|^2 \geq 0.$$

$\square$

Remark 2.1. If $H > 1$, we can show that the covariance function C given by (2.1) is not positive definite. Taske $t_1 = 1, t_2 = 2, \lambda_1 = -2, \lambda_2 = 1$. Then

$$\sum_{i,j=1}^{2} \lambda_i \lambda_j C(t_i, t_j) = 4 - 2^{2H} < 0.$$

This example is taken from [Nourdin (2012)].

Now we introduce the fractional Brownian motion (fBm for short).

Definition 2.1. Let $T \subset \mathbb{R}$. The fractional Brownian motion with Hurst parameter $H \in (0, 1]$ is defined as a centered Gaussian process $(B_t^H, t \in T)$ with covariance function

$$\mathbf{E} B_t^H B_s^H = \frac{1}{2} \left(|t|^{2H} + |s|^{2H} - |t - s|^{2H} \right) := R_H(t, s) \qquad (2.6)$$

for every $s, t \in T$.

Proposition 2.1 guarantees the existence of the fractional Brownian motion. If $T = [0, \infty)$, the centered Gaussian process $(B_t^H, t \geq 0)$ with covariance given by (2.6) will be called *standard fractional Brownian motion*. If $H = \frac{1}{2}$ it reduces to the standard Brownian motion (or the standard Wiener process), which is a centered Gaussian process $(B_t, t \geq 0)$ with

$$\mathbf{E} B_t B_s = t \wedge s, \quad t, s \geq 0.$$

In particular, the standard Brownian motion has independent increments, i.e. $B_{t_1}, B_{t_2} - B_{t_1}, \ldots, B_{t_n} - B_{t_{n-1}}$ are independent random variables for every $n \geq 1$ and for every $0 \leq t_1 < t_2 \ldots < t_n$.

If $H = 1$, then the standard fBm reduces to the trivial process $(tZ)_{t \geq 0}$ with $Z \sim N(0, 1)$. Indeed,

$$\mathbf{E}(tZsZ) = ts\mathbf{E}Z^2 = ts, \quad s, t \geq 0.$$

Remark 2.2.

(1) If $H = \frac{1}{2}$ and $T = \mathbb{R}$, then the process from Definition 2.1 is the two-sided Brownian motion, which is a centered Gaussian process $(W_t, t \in \mathbb{R})$ with covariance

$$\mathbf{E}W_t W_s = \frac{1}{2}\left(|t| + |s| - |t - s|\right), \quad s, t \in \mathbb{R}. \tag{2.7}$$

There is an alternative way to define the two-sided Wiener process. Consider two independent (standard) Brownian motions $(W_t^{(1)}, t \geq 0)$ and $(W_t^{(2)}, t \geq 0)$. We set $(W_t, t \in \mathbb{R})$ by

$$W_t = \begin{cases} W_t^{(1)} & \text{for } t > 0 \\ W_{-t}^{(2)} & \text{for } t < 0 \end{cases} \tag{2.8}$$

with $W_0 = 0$. Then for every $s, t \in \mathbb{R}$,

$$\mathbf{E}W_t W_s = \begin{cases} t \wedge s, & \text{if } s, t \geq 0 \\ 0 & \text{if } t < 0 \text{ or } s < 0 \\ (-t) \wedge (-s) & \text{if } s, t < 0 \end{cases} = \frac{1}{2}\left(|t| + |s| - |t - s|\right). \tag{2.9}$$

Hence $(W_t, t \in \mathbb{R})$ is a centered Gaussian process with covariance (2.7).

(2) We cannot define the fBm $(B_t^H, t \geq 0)$ on the whole real line as we did for the Wiener process, via the formula (2.8). Consider two independent standard fractional Brownian motions $(B_t^{1,H}, t \geq 0)$ and $(B_t^{2,H}, t \geq 0)$ with the same Hurst parameter $H \in (0, 1)$. Set

$$X_t^H = \begin{cases} B_t^{1,H}, & \text{for } t \geq 0 \\ B_{-t}^{2,H}, & \text{for } t < 0. \end{cases} \tag{2.10}$$

The process $(X_t^H, t \geq 0)$ is a centered Gaussian process but it is not a fBm with time interval $\mathbb{R}$ if $H \neq \frac{1}{2}$. Indeed, if $t > 0$ and $s < 0$, then

$$\mathbf{E}X_t^H X_s^H = 0 \neq \frac{1}{2}\left(|t|^{2H} + |s|^{2H} - |t - s|^{2H}\right)$$

$$= \frac{1}{2}\left(t^{2H} + (-s)^{2H} - (t - s)^{2H}\right).$$

Sometimes in the literature, the process $(X_t^H, t \geq 0)$ given by (2.10) is still called two-sided fractional Brownian motion.

We discuss the main properties of the fBm.

Proposition 2.2. *Let* $T = \mathbb{R}_+$ *or* $T = \mathbb{R}$. *Let* $(B_t^H, t \in T)$ *be a fractional Brownian motion with* $H \in (0, 1)$. *Then* B^H *is a* H-*self-similar stochastic process.*

Proof. Recall the Definition 1.4 and the comments that follows it. Let $c > 0, T = \mathbb{R}_+$ (the same arguments will hold for $T = \mathbb{R}$). The processes $(B_{ct}^H, t \geq 0)$ and $(c^H B_t^H, t \geq 0)$ are both Gaussian and centered with the same covariance given by

$$c^{2H} \frac{1}{2} \left(t^{2H} + s^{2H} - |t - s|^{2H} \right), \quad s, t \geq 0.$$

By Theorem 1.4, the two processes have the same finite-dimensional distributions. □

Proposition 2.3. *Let $T = \mathbb{R}_+$ or $T = \mathbb{R}$. The fractional Brownian motion $(B_t^H, t \in T)$ has stationary increments.*

Proof. Assume $T = \mathbb{R}_+$ and let us show that for every $h > 0$, the stochastic process $\left(B_{t+h}^H - B_h^H, t \geq 0 \right)$ has the same finite dimensional distributions as $(B_t^H)_{t \geq 0}$. Let

$$X_t = B_{t+h}^H - B_h^H, \quad t \geq 0.$$

The process $(X_t, t \geq 0)$ is Gaussian, centered, with the covariance

$$
\begin{aligned}
&\mathbf{E} X_t X_s \\
&= \mathbf{E} \left((B_{t+h}^H - B_h^H)(B_{s+h}^H - B_h^H) \right) \\
&= \frac{1}{2} \left[((t+h)^{2H} + (s+h)^{2H} - |t-s|^{2H}) - ((t+h)^{2H} + h^{2H} - |t|^{2H}) \right. \\
&\quad \left. - ((s+h)^{2H} + h^{2H} - |s|^{2H}) + 2h^{2H} \right] \\
&= \frac{1}{2}(t^{2H} + s^{2H} - |t-s|^{2H}),
\end{aligned}
$$

which is the covariance in (2.6). So it has the same finite dimensional distributions as B^H. □

Remark 2.3. Propositions 2.2 and 2.3 suggest an alternative definition of the fractional Brownian motion $(B_t^H)_{t \in T}$ with $T = \mathbb{R}_+$ or $T = \mathbb{R}$. This process can be defined as the only self-similar Gaussian process with stationary increments and $\mathbf{E}(B_1^H)^2 = 1$. Under these hypothesis, we obtain by Theorem 1.2 that such a process has the covariance of the fBm.

Concerning the regularity of the sample paths of the fractional Brownian motion, we have the following result.

Proposition 2.4. *Let $B^H = (B_t^H, t \in T)$ with $T \subset \mathbb{R}$, be a fBm with Hurst parameter $H \in (0,1)$. Then B^H admits a version whose trajectories are Hölder continuous of order δ for any $\delta \in (0, H)$.*

Proof. By the covariance formula (2.6), for $s, t \in T$,

$$\mathbf{E}|B_t^H - B_s^H|^2$$
$$= |t|^{2H} + |s|^{2H} - 2\frac{1}{2}\left(|t|^{2H} + |s|^{2H} - |t-s|^{2H}\right)$$
$$= |t-s|^{2H}$$

so, for every $s, t \in T$, with $s \leq t$

$$B_t^H - B_s^H \sim N(0, (t-s)^{2H}).$$

This implies, for every $p \geq 1$,

$$\mathbf{E}|B_t^H - B_s^H|^p = \mathbf{E}|Z||t-s|^{Hp}$$

with $Z \sim N(0,1)$. We can apply the Kolmogorov continuity theorem (Theorem 1.1) for $p > \frac{1}{H}$ large enough to get the Hölder continuity of order $\delta \in (0, H)$. $\qquad\square$

Definition 2.2. If $B^H = (B_t^H, t \in \mathbb{R})$ is a fBm with Hurst parameter $H \in (0,1)$, we define the fractional Gaussian noise as the discrete stochastic process $(X_n, n \in \mathbb{Z})$ given by

$$X_n = B_{n+1}^H - B_n^H \text{ for every integer } n \in \mathbb{Z}.$$

This random sequence is stationary, i.e. for every $k \in \mathbb{Z}$, the finite dimensional distributions of the discrete process $(X_{n+k}, n \in \mathbb{Z})$ do not depend on k. Indeed, for every $n, m \geq 0$,

$$\mathbf{E}X_{n+k}X_{m+k} = \frac{1}{2}\left(|n-m+1|^{2H} + |n-m-1|^{2H} - 2|n-m|^{2H}\right)$$
$$= \varphi(n-m).$$

We denoted by

$$\varphi(v) = \frac{1}{2}\left(|v+1|^{2H} + |v-1|^{2H} - 2|v|^{2H}\right), \quad v \in \mathbb{Z}. \tag{2.11}$$

This is called the auto-correlation function of the fractional Gaussian noise $(X_n, n \in \mathbb{Z})$ and it characterizes the law of this sequence. In particular,

$$\varphi_0 = 0, \quad \varphi(-v) = \varphi(v)$$

and for $|n| \geq 1$

$$\varphi(n) = \mathbf{E}X_nX_0 = \mathbf{E}(B_{n+1}^H - B_n^H)B_1^H.$$

For $H = \frac{1}{2}$, we can see

$$\varphi(v) = 0 \text{ for every } v \in \mathbb{Z}.$$

As a consequence, we have the following result.

Proposition 2.5. *Let φ be given by (2.11). Then*

$$\sum_{v \in \mathbb{Z}} \varphi(v) = \infty \text{ if } H > \frac{1}{2} \tag{2.12}$$

and

$$\sum_{v \in \mathbb{Z}} \varphi(v) < \infty \text{ if } H \leq \frac{1}{2}. \tag{2.13}$$

Proof. For $H = \frac{1}{2}$, the result is obvious. Let $H \neq \frac{1}{2}$. The conclusion comes from the fact that $\varphi(v)$ behaves, for $|v| \to \infty$ as

$$H(2H - 1)|v|^{2H-2}.$$

$\square$

Remark 2.4. One usually interpret (2.12)-(2.13) by saying that the fBm has long memory (or long-range dependence) for $H > \frac{1}{2}$ and short memory if $H \leq \frac{1}{2}$. More generally, a centered stationary sequence $(X_n, n \in \mathbb{Z})$ is called long-range dependent if

$$\sum_{n \in \mathbb{Z}} \mathbf{E}(X_1 X_n) = \infty.$$

If we above series converges, we say that the sequence is short-range dependent.

The standard fractional Brownian motion also satisfies the following time inversion property.

Proposition 2.6. *Let $(B_t^H, t \geq 0)$ be a standard fBm. Then the processes $\left(t^{2H} B_{\frac{1}{t}}^H, t > 0\right)$ and $(B_t^H, t > 0)$ have the same finite dimensional distributions.*

Proof. Both processes from the statement are centered Gaussian processes and for $s, t > 0$ arbitrary,

$$\mathbf{E}\left(t^{2H} B_{\frac{1}{t}}^H s^{2H} B_{\frac{1}{s}}^H\right) = \frac{1}{2} t^{2H} s^{2H} \left[\left(\frac{1}{t}\right)^{2H} + \left(\frac{1}{s}\right)^{2H} - \left|\frac{1}{t} - \frac{1}{s}\right|^{2H}\right]$$

$$= R_H(t, s)$$

where R_H is the covariance of fBm (2.6). $\square$

Recall that $(X_t, t \geq 0)$ is a Markov process if for every $A \in \mathcal{B}(\mathbb{R})$ and for every $0 \leq s \leq t$,

$$P(X_t \in A/\mathcal{F}_s) = P(X_t \in A/\sigma(X_s)),$$

where $\mathcal{F}_s = \sigma(X_u, 0 \leq u \leq s)$ is the filtration generated by the process X.

Proposition 2.7. *The standard fractional Brownian motion is not a Markov process unless its Hurst index H is equal to $\frac{1}{2}$.*

Proof. Recall that (see [Revuz and Yor (1994)]) a Gaussian process with covariance R is Markovian if and only if

$$R(s, u)R(t, t) = R(s, t)R(t, u)$$

for every $s \leq t \leq u$. One can prove (see e.g. [Nourdin (2012)], Proposition 2.3) that B^H does not satisfy this condition if $H \neq \frac{1}{2}$. $\qquad \square$

Assume B^H is a fBm with $H \in (0, 1)$. Let $0 \leq A_1 < A_2$ two real numbers and consider

$$t_i = A_1 + \frac{i}{N}(A_2 - A_1) \quad , \tag{2.14}$$

a partition of the interval $[A_1.A_2]$. Define, for $p \geq 1$,

$$S^{N,p}_{[A_1,A_2]}(B^H) = \sum_{i=0}^{N-1} \left| B^H_{t_{i+1}} - B^H_{t_i} \right|^p. \tag{2.15}$$

This is called the p-variation sequence of the fractional Brownian motion over the interval $[A_1, A_2]$. If $[A_1, A_2] = [0, t]$ with $t \geq 0$, we simply denote

$$S^{N,p}_{[0,t]}(B^H) = V^{N,p}_t(B^H).$$

The behavior of the sequence $S^{N,p}_{[A_1,A_2]}(B^H)$ plays an important role to the parameter estimation for the stochastic partial differential equations discussed in the next chapters.

We recall the following fact: if $(X_j)_{j \geq 0}$ is a stationary centered Gaussian sequence such that

$$\mathbf{E}X_0 X_j \to_{j \to \infty} 0$$

then the sequence $(X_j)_{j \geq 0}$ is *ergodic*, i.e. it satisfies

$$\frac{1}{N} \sum_{j=1}^{N} f(X_j) \to_{N \to \infty} \mathbf{E}f(X_1) \text{ almost surely and in } L^1(\Omega) \tag{2.16}$$

for every measurable function $f : \mathbb{R} \to \mathbb{R}$ such that $\mathbf{E}|f(X_1)| < \infty$, see e.g. [Samorodnitsky (2016)].

Let us state and prove the following result.

Theorem 2.1. *Let $(B_t^H, t \geq 0)$ be a fBm with Hurst index $H \in (0,1)$ and let $\left(S_{[A_1,A_2]}^{N,p}(B^H), N \geq 1 \right)$ be given by (2.15). Then*

$$S_{[A_1,A_2]}^{N,p}(B^H) \to_{N\to\infty} \begin{cases} 0, & \text{if } p > \frac{1}{H} \\ |A_2 - A_1|\mathbf{E}|Z|^{\frac{1}{H}}, & \text{if } p = \frac{1}{H} \\ +\infty, & \text{if } p < \frac{1}{H} \end{cases} \qquad (2.17)$$

where $Z \sim N(0,1)$.

Proof. Let $p \geq 1$. Consider the sequence $(Y^{N,p}, N \geq 1)$

$$Y^{N,p} = N^{pH-1} V_{[A_1,A_2]}^{N,p}(B^H) = N^{pH-1} \sum_{i=0}^{N-1} \left| B_{t_{i+1}}^H - B_{t_i}^H \right|^p. \qquad (2.18)$$

By the self-similarity and the stationarity of the increments of the fBm (Propositions 2.2 and 2.3), we have the following equality in distribution

$$Y^{N,p} =^{(d)} |A_2 - A_1|^{Hp} \frac{1}{N} \sum_{i=0}^{N-1} \left| B_{i+1}^H - B_i^H \right|^p := U^{N,p}. \qquad (2.19)$$

Let $X_j = B_{j+1}^H - B_j^H$ for $j \geq 0$. Then

$$\mathbf{E} X_j X_0 = \mathbf{E}(B_{j+1}^H - B_j^H) B_1^H$$
$$= \frac{1}{2} \left(|j+1|^{2H} + |j-1|^{2H} - 2|j|^{2H} \right)$$
$$\sim H(2H-1) j^{2H-2} \to_{j\to\infty} 0.$$

thus the sequence $(X_j, j \geq 0)$ is ergodic and by (2.16),

$$U^{N,p} \to_{N\to\infty} |A_2 - A_1|^{Hp} \mathbf{E} |B_1^H|^p = |A_2 - A_1|^{Hp} \mathbf{E} |Z|^p$$

almost surely and in $L^1(\Omega)$. By (2.19), the sequence $Y^{N,p}$ converges in law as $N \to \infty$, to $|A_2 - A_1|^{Hp} \mathbf{E}|Z|^p$. Therefore, since the convergence in law to a constant implies the convergence in probability, we obtain that $Y^{N,p}$ converges in probability as $N \to \infty$, to $|A_2 - A_1|^{Hp} \mathbf{E}|Z|^p$. This implies the conclusion (2.17). $\square$

Proposition 2.8. *Let B^H be a fBm. Then B^H is not a semimartingale, unless $H = \frac{1}{2}$.*

Proof. From (2.17), with $p = 2$, we get that for every $t \geq 0$

$$S_t^{N,2}(B^H) \to_{N \to \infty} \begin{cases} 0, & \text{if } H > \frac{1}{2} \\ +\infty, & \text{if } H < \frac{1}{2} \text{ in probability.} \end{cases}$$

This implies that the fBm cannot be a semi-martingale if $H \neq \frac{1}{2}$ (see e.g. [Karatzas and Shreve (1988)]). $\qquad\square$

2.2 Bifractional Brownian motion

The bifractional Brownian motion represents an extension of the fractional Brownian motion. It depends on two Hurst parameters and this gives more flexibility to model certain phenomena. It is also strongly connected with the solution to the stochastic heat equation, as we will see later.

We start by showing the the function that defines the covariance of the bifractional Brownian motion is positive definite.

Proposition 2.9. *Let $H \in (0,1)$ and $K \in (0,1]$. Consider the function $R_{H,K} : \mathbb{R}^2 \to \mathbb{R}$,*

$$R_{H,K}(t,s) = \frac{1}{2^K} \left((t^{2H} + s^{2H})^K - |t - s|^{2HK} \right), \qquad s, t \in \mathbb{R}.$$

Then the function $R_{H,K}$ is symmetric and of positive type.

Proof. Assume $K \in (0,1)$ (the case $K = 1$ is already known from Proposition 2.1). Take $N \geq 1$, $\lambda_1, \ldots, \lambda_N \in \mathbb{R}$ and $t_1, \ldots, t_N \in \mathbb{R}$ arbitrary. We have, by using (2.2),

$$\sum_{i,j=1}^{N} \lambda_i \lambda_j R_{H,K}(t_i, t_j)$$

$$= 2^{-K} \frac{K}{\Gamma(1-K)} \int_0^\infty \sum_{i,j=1}^{N} \lambda_i \lambda_j e^{-x(|t_i|^{2H} + |t_j|^{2H})}$$

$$\times \left(e^{x(|t_i|^{2H} + |t_j|^{2H} - |t_i - t_j|^{2H})} - 1 \right) x^{-1-K} dx.$$

Since the function $A(t,s) = |t|^{2H} + |s|^{2H} - |t - s|^{2H}$ is of positive type if $H \in (0,1)$, so is the function $C(t,s) = e^{A(t,s)} - 1$ (see Remark 1.3). Set

$$c_{i,x} = \lambda_i e^{-x|t_i|^{2H}} \text{ for } i = 1, \ldots, N.$$

Then we have

$$\sum_{i,j=1}^{N} \lambda_i \lambda_j R_{H,K}(t_i, t_j)$$

$$= 2^{-K} \frac{K}{\Gamma(1-K)} \int_0^\infty \sum_{i,j=1}^{N} c_{i,x} c_{j,x} C(x^{\frac{1}{2H}} t_i, x^{\frac{1}{2H}} t_j) x^{-1-K} dx \geq 0.$$

The symmetry of the function $R_{H,K}$ is obvious and the conclusion is obtained. $\qquad \square$

We now introduce the bifractional Brownian motion.

Definition 2.3.
Let $H \in (0,1)$ and $K \in (0,1]$ and $T \subset \mathbb{R}$. The bifractional Brownian motion with Hurst parameter H and K is defined as a centered Gaussian process $(B_t^{H,K}, t \in T)$ with covariance

$$\mathbf{E} B_t^{H,K} B_s^{H,K} = R_{H,K}(t,s) = \frac{1}{2^K} \left((t^{2H} + s^{2H})^K - |t-s|^{2HK} \right) \quad (2.20)$$

for every $s, t \in T$.

Proposition 2.9 gives the existence of the bifractional Brownian motion (bi-fBm in short). If $K = 1$, then $B^{H,1}$ is nothing else than the fractional Brownian motion with Hurst parameter $H \in (0,1)$. Let us first notice that $B^{H,K}$ is a self-similar process.

Proposition 2.10. *Let $T = \mathbb{R}_+$ or $T = \mathbb{R}$. Then $(B_t^{H,K}, t \in T)$ is HK-self-similar.*

Proof. For $c > 0$, the centered Gaussian processes

$$(B_{ct}^{H,K}, t \in T) \text{ and } (c^{HK} B_t^{H,K}, t \in T)$$

have the same covariance given by

$$c^{2HK} R_{H,K}(t,s), \quad t, s \in T$$

with $R_{H,K}$ from (2.20). $\qquad \square$

The bi-fBm cannot have stationary increments (otherwise, it would coincide with the fBm, see Remark 2.3). But we can prove the below property for its increments. The following bound of the mean square of the increment will play an important role.

Proposition 2.11. *Let $H \in (0,1), K \in (0,1]$ and let $(B_t^{H,K}, t \in \mathbb{R})$ be a bifractional Brownian motion. Then for every $s,t \in \mathbb{R}$,*

$$2^{-K}|t-s|^{2HK} \leq \mathbf{E}\left(B_t^{H,K} - B_s^{H,K}\right)^2 \leq 2^{1-K}|t-s|^{2HK}. \qquad (2.21)$$

Proof. Let us prove the lower bound. From (2.20), we have

$$\mathbf{E}\left|B_t^{H,K} - B_s^{H,K}\right|^2 = 2^{1-K}|t-s|^{2HK} \qquad (2.22)$$
$$+ \left(|t|^{2HK} + |s|^{2HK} - 2^{1-K}(|t|^{2H} + |s|^{2H})^K\right), s,t \in \mathbb{R}$$

so to obtain the lower bound in (2.21) is equivalent to show that

$$|t-s|^{2HK} + 2^K(|t|^{2HK} + |s|^{2HK}) - 2(|t|^{2H} + |s|^{2H})^K \geq 0 \qquad (2.23)$$

for every $s,t \in \mathbb{R}$. The above relation is clearly true if $s = 0$ or $t = 0$. Let us treat the other cases.

Case $0 < s \leq t$. We can write

$$|t-s|^{2HK} + 2^K(t^{2HK} + s^{2HK}) - 2(t^{2H} + s^{2H})^K = s^{2HK} f\left(\frac{t}{s}\right)$$

with, for $x \geq 1$

$$f(x) = |x-1|^{2HK} + 2^K - +2^K|x|^{2HK} - 2(1 + |x|^{2H})^K$$
$$= (x-1)^{2HK} + 2^K - +2^K x^{2HK} - 2(1 + x^{2H})^K. \qquad (2.24)$$

We need to show that $f(x) \geq 0$ for every $x \geq 1$. We differentiate f and we obtain

$$f'(x) = 2HK x^{2HK-1}\left[\left(1 - \frac{1}{x}\right)^{2HK-1} + 2^K - 2\left(1 + \frac{1}{x^{2H}}\right)^{K-1}\right].$$

For $2HK \leq 1$, we can easily see that $f'(x) \geq 0$ for every $x \geq 1$ and noting that $f(1) = 0$, we get the conclusion. If $1 < 2HK < 2$ then $2H \geq 1$ and $K \geq \frac{1}{2}$, then we also have $f'(x) \geq 0$ for every $x \geq 1$ by differentiating once again the function

$$\left(1 - \frac{1}{x}\right)^{2HK-1} + 2^K - 2\left(1 + \frac{1}{x^{2H}}\right)^{K-1}.$$

Case $s \leq t < 0$. We can write (2.23) as

$$|t|^{2HK} f\left(\frac{s}{t}\right)$$

and this is true due to the above computations, since $\frac{s}{t} \geq 1$.

Case $s < 0 < t$. When $t \geq -s$ (or equivalently $\frac{t}{s} \leq -1$), relation (2.23) is equivalent to showing

$$|s|^{2HK} f\left(\frac{t}{s}\right) \geq 0 \text{ with } \frac{t}{s} \leq -1.$$

If $t < -s$ (or $\frac{s}{t} \leq -1$), the inequality (2.23) is equivalent to

$$|t|^{2HK} f\left(\frac{s}{t}\right) \geq 0 \text{ with } \frac{s}{t} \leq -1.$$

Therefore, this case will be concluded if we show that

$$f(x) \geq 0 \text{ for } x \leq -1$$

with f from (2.24). Now, by setting $v = -x$, this is the same as proving

$$g(v) = (1 + v) + 2^K + 2^K v^{2HK} - 2(1 + v^{2H})^K \geq 0 \text{ for } v \geq 1. \qquad (2.25)$$

On the other hand, it is easy to see that $g(v) \geq f(v)$ for every $v \geq 1$ and thus (2.25) holds true.

Let us prove the upper bound in (2.21). This follows from (2.22) by observing that the quantity

$$\left(|t|^{2HK} + |s|^{2HK} - 2^{1-K}(|t|^{2H} + |s|^{2H})^K\right)$$

is positive for every $s, t \mathbb{R}$. Indeed, the function $h(x) = x^K$ is concave for $x > 0$ and we use

$$h\left(\frac{|t|^{2H} + |s|^{2H}}{2}\right) \geq \frac{h(|t|^{2H}) + h(|s|^{2H})}{2}.$$

$\square$

Define, for $0 < K < 1$, the process

$$X_t^K = \int_0^\infty \left(1 - e^{-\theta t}\right) \theta^{-\frac{1+K}{2}} dW_\theta \qquad (2.26)$$

where $(W_\theta, \theta \in \mathbb{R}_+)$ is a Wiener process and the above Wiener integral is presented in Section 4.2. Then X^K is a centered Gaussian process with covariance

$$\mathbf{E} X_t^K X_s^K := R^X(t, s) = \int_0^\infty \left(1 - e^{-\theta t}\right)\left(1 - e^{-\theta s}\right) \theta^{-1-K} d\theta \qquad (2.27)$$

$$= \frac{\Gamma(1 - K)}{K}(t^K + s^K - (t + s)^K)$$

for every $s, t \geq 0$. The last equality above comes from (2.2).

Theorem 2.2. *Let* $(B_t^{H,K}, t \geq 0)$ *be a bi-fBm and consider* $(W_\theta, \theta \geq 0)$ *a Wiener process independent of* $B^{H,K}$. *Define for every* $t \geq 0$

$$X_t^{H,K} := X_{t^{2H}}^{K}. \tag{2.28}$$

Then the processes

$$\left(C_1 X_t^{H,K} + B_t^{H,K}, t \geq 0\right) \text{ and } (C_2 B_t^{HK}, t \geq 0)$$

have the same law, where $C_1 = \sqrt{\frac{K 2^{-K}}{\Gamma(1-K)}}$ *and* $C_2 = 2^{\frac{1-K}{2}}$.

Proof. Denote by

$$Y_t^{H,K} = C_1 X_t^{H,K} + B_t^{H,K}$$

for every $t \geq 0$. Then by (2.27), for every $s, t \geq 0$, via the independence of $X^{H,K}$ and $B^{H,K}$,

$$\begin{aligned}
\mathbf{E} Y_t^{H,K} Y_s^{H,K} &= C_1^2 \mathbf{E} X_t^{H,K} X_s^{H,K} + \mathbf{E} B_t^{H,K} B_s^{H,K} \\
&= 2^{-K} (t^{2HK} + s^{2HK} - (t^{2H} + s^{2H})^K) \\
&\quad + 2^{-K} ((t^{2H} + s^{2H})^K - |t - s|^{2HK}) \\
&= 2^{-K} (t^{2HK} + s^{2HK} - |t - s|^{2HK}) \\
&= \mathbf{E} C_2^2 B_t^{HK} B_s^{HK}.
\end{aligned}$$

$\square$

An important fact is that the trajectories of the process X^K are very regular.

Proposition 2.12. *Consider the process* X^K *given by (2.26). Then the trajectories of* X^K *are (modulo a modification) absolutely continuous and infinitely differentiable on* $(0, \infty)$.

Proof. Note that we have, by Fubini, for every $t > 0$,

$$X_t^K = \int_0^t Y_s \, ds$$

with

$$Y_t = \int_0^\infty \theta^{\frac{1-K}{2}} e^{-\theta t} dW_\theta, \quad t \geq 0.$$

We have $\mathbf{E} Y_t^2 = C t^{K-2}$ with $C > 0$ and thus

$$\mathbf{E} \int_0^t |Y_s| ds \leq C \int_0^t \left(\mathbf{E} |Y_s|^2 \right)^{\frac{1}{2}} ds = C \int_0^t s^{\frac{K-2}{2}} ds < \infty$$

since $K > 0$. The fundamental theorem of Lebesque integral calculus implies that the paths of X^K are almost surely absolutely continuous and

$$\frac{dX_t^K}{dt} = Y_t \text{ for every } t > 0.$$

Analogously, we get that the n th derivative of X_t^K

$$\frac{d^n}{dt^n} X_t^K = \int_0^\infty (-1)^{n-1} \theta^{n-\frac{1}{2}-\frac{K}{2}} e^{-\theta s} dW_\theta$$

is well-defined. □

Remark 2.5. The above result provides another argument of the upper inequality in (2.21). From the decomposition in Theorem 2.2, we have for every $s, t \geq 0$,

$$\mathbf{E}|B_t^{H,K} - B_s^{H,K}|^2 + C_1^2 \mathbf{E}\left|X_t^{H,K} - X_s^{H,k}\right|^2 = 2^{1-K} \mathbf{E}\left|B_t^{HK} - B_s^{HK}\right|^2,$$

and this implies

$$\mathbf{E}|B_t^{H,K} - B_s^{H,K}|^2 = 2^{1-K} \mathbf{E}\left|B_t^{HK} - B_s^{HK}\right|^2 - C_1^2 \mathbf{E}\left|X_t^{H,K} - X_s^{H,k}\right|^2$$

$$\leq 2^{1-K} \mathbf{E}\left|B_t^{HK} - B_s^{HK}\right|^2 \leq 2^{1-K}|t-s|^{2HK}$$

which represents the upper bound in (2.21).

We can deduce the behavior of the p-variations of the bifractional Brownian motion. If $(t_i, i = 0, 1, \ldots, N)$ is the partition of $[A_1, A_2]$ given by (2.14), then we set for $p \geq 1$

$$S_{[A_1,A_2]}^{N,p}(B^{H,K}) = \sum_{i=0}^{N-1} \left|B_{t_{i+1}}^{H,K} - B_{t_i}^{H,K}\right|^p. \tag{2.29}$$

When $K = 1$, this is the quadratic variation of the fBm, see (2.15). If $[A_1, A_2] = [0, t]$, we still use the notation $S_{[0,t]}^{N,p}(B^{H,K}) = S_t^{N,p}(B^{H,K})$.

Theorem 2.3. *Let $(B_t^{H,K}, t \geq 0)$ be a bi-fBm with $H \in (0,1), K \in (0,1]$. Then*

$$S_{[A_1,A_2]}^{N,p}(B^{H,K}) \to_{N\to\infty} \begin{cases} 0, & \text{if } p > \frac{1}{HK} \\ 2^{\frac{1-K}{2}} |A_2 - A_1| \mathbf{E}|Z|^{\frac{1}{HK}}, & \text{if } p = \frac{1}{HK} \\ +\infty, & \text{if } p < \frac{1}{HK}. \end{cases} \tag{2.30}$$

Proof. By using the decomposition in Theorem 2.2 and the Minkowski inequality

$$\left(\sum_{i=0}^{n-1} |x_i|^p\right)^{\frac{1}{p}} - \left(\sum_{i=0}^{n-1} |y_i|^p\right)^{\frac{1}{p}} \le \left(\sum_{i=0}^{n-1} |x_i - y_1|^p\right)^{\frac{1}{p}}$$

$$\le \left(\sum_{i=0}^{n-1} |x_i|^p\right)^{\frac{1}{p}} + \left(\sum_{i=0}^{n-1} |y_i|^p\right)^{\frac{1}{p}} \tag{2.31}$$

we can write, with $C_1, C_2 = 2^{\frac{1-K}{2}}$ from Theorem 2.2,

$$C_2 \left(\sum_{i=0}^{n-1} |B_{t_{i+1}}^{HK} - B_{t_i}^{HK}|^p\right)^{\frac{1}{p}} - C_1 \left(\sum_{i=0}^{n-1} |X_{t_{i+1}}^{H,K} - X_{t_i}^{HK}|^p\right)^{\frac{1}{p}}$$

$$\le \left(\sum_{i=0}^{n-1} |B_{t_{i+1}}^{H,K} - B_{t_i}^{H,K}|^p\right)^{\frac{1}{p}}$$

$$\le C_2 \left(\sum_{i=0}^{n-1} |B_{t_{i+1}}^{HK} - B_{t_i}^{HK}|^p\right)^{\frac{1}{p}} + C_1 \left(\sum_{i=0}^{n-1} |X_{t_{i+1}}^{H,K} - X_{t_i}^{HK}|^p\right)^{\frac{1}{p}}$$

where B^{HK} is a fBm with Hurst index HK and $X^{H,K}$ is a process with absolute continuous paths given by (2.28).

Let $p = \frac{1}{HK}$. By Theorem 2.1, the sequence

$$2^{\frac{1-K}{2}} \left(\sum_{i=0}^{n-1} |B_{t_{i+1}}^{HK} - B_{t_i}^{HK}|^{\frac{1}{HK}}\right)$$

converges to the desired limit (the right-hand side of (2.30)). It remains to show that the sequence

$$\sum_{i=0}^{n-1} |X_{t_{i+1}}^{H,K} - X_{t_i}^{HK}|^{\frac{1}{HK}}$$

converges to zero as $N \to \infty$. We have

$$\sum_{i=0}^{n-1} |X_{t_{i+1}}^{H,K} - X_{t_i}^{HK}|^{\frac{1}{HK}}$$

$$\le \sup_{i=0,\ldots,N} |X_{t_{i+1}}^{H,K} - X_{t_i}^{H,K}|^{\frac{1}{HK}-1} \sum_{i=0}^{n-1} |X_{t_{i+1}}^{H,K} - X_{t_i}^{HK}|$$

and the absolute continuity of the paths of $X^{H,K}$ implies that

$$\sup_{i=0,\ldots,N} |X_{t_{i+1}}^{H,K} - X_{t_i}^{H,K}|^{\frac{1}{HK}-1} \to_{N\to\infty} 0$$

and

$$\sum_{i=0}^{n-1} |X_{t_{i+1}}^{H,K} - X_{t_i}^{HK}| < \infty.$$

If $p > \frac{1}{HK}$, then $V_{[A_1,A_2]}^{N,p}(B^{H,K})$ converges to zero in $L^1(\Omega)$ due to the upper bound in (2.21). Indeed,

$$\mathbf{E} \sum_{i=0}^{n-1} |B_{t_{i+1}}^{H,K} - B_{t_i}^{H,K}|^p \leq C \sum_{i=0}^{N-1} |t_{i+1} - t_i|^{HKp}$$

$$\leq C \sup_{i=0,\dots,N} |t_{i+1} - t_i|^{HKp-1} \sum_{i=0}^{N-1} |t_{i+1} - t_i|$$

$$\to_{N\to\infty} 0.$$

The lower bound in (2.21) will imply the result when $p < \frac{1}{HK}$. $\qquad\square$

Remark 2.6. The convergence (2.30) shows that $B^{H,K}$ cannot be a semi-martingale if $2HK \neq 1$. If suffices to use the argument in the proof of Proposition 2.8.

Chapter 3

Multiparameter Gaussian processes

The stochastic partial differential equations discussed later will be driven by multiparameter Gaussian process (or Gaussian sheets). The purpose of this chapter is to present the definition and basic properties of some important Gaussian sheets.

3.1 The anisotropic fractional Brownian sheet

There are several ways to define a multiparameter fractional Brownian motion. The first approach is the following.

Definition 3.1. Let $T \subset \mathbb{R}^d$. The d-parameter anisotropic fractional Brownian motion (or the d-parameter anisotropic fractional Brownian sheet) with Hurst index $H = (H_1, \ldots, H_d) \in (0, 1)^d$ is defined as a centered Gaussian process $(W_t^H, t \in T)$ with covariance function

$$\mathbf{E}W_t^H W_s^H = \prod_{j=1}^{d} \left(\frac{1}{2}(|t_j|^{2H_j} + |s_j|^{2H_j} - |t_j - s_j|^{2H_j}) \right)$$
$$:= R_H(t, s), \tag{3.1}$$

for every $t = (t_1, \ldots, t_d), s = (s_1, \ldots, s_d) \in T$.

Remark 3.1. Since the product of two covariance functions is still a covariance function (see Remark 1.3), the right-hand side of (3.1) defines a covariance function.

The standard anisotropic fractional Brownian sheet is a centered Gaussian process $(B_t^H, t \in [0, \infty)^d)$ with covariance given by (3.1). If $H_1 = H_2 = \ldots = H_d = \frac{1}{2}$, then the process $W = (W_t^{\frac{1}{2}}, t \in [0, \infty)^d)$ is called the standard anisotropic d-parameter Brownian sheet. It is defined as a centered Gaussian process $(W_t, t \in [0, \infty)^d)$ with covariance function

$$\mathbf{E} W_t W_s = t \wedge s = \prod_{j=1}^{d} (t_j \wedge s_j)$$

for every $t = (t_1, \ldots, t_d), s = (s_1, \ldots, s_d) \in [0, \infty)^d$.

In the sequel we use the following notation: For $d \in \mathbb{N} \backslash \{0\}$ if $a = (a_1, a_2, \ldots, a_d)$, $b = (b_1, b_2, \ldots, b_d)$, $\alpha = (\alpha_1, \ldots, \alpha_d)$ are vectors in $\mathbb{R}^d$, we set

$$a = \prod_{i=1}^{d} a_i b_i, \quad |a - b|^\alpha = \prod_{i=1}^{d} |a_1 - b_1|^{\alpha_i},$$

$$[a, b] = \prod_{i=1}^{d} [a_i, b_i], \quad (a, b) = \prod_{i=1}^{d} (a_i, b_i),$$

$$a/b = (a_1/b_1, a_2/b_2, \ldots, a_d/b_d), \quad a^b = \prod_{i=1}^{d} a_i^{b_i} \qquad (3.2)$$

and $a < b$ if $a_1 < b_1, a_2 < b_2, \ldots, a_d < b_d$ (analogously for the other inequalities).

Let us recall the concept of self-similarity for multiparameter stochastic processes.

Definition 3.2. Let $T = \mathbb{R}^d$ or $T = \mathbb{R}_+^d$. A stochastic process $(X_t, t \in T)$, is called self-similar with self-similarity order $\alpha = (\alpha_1, \ldots, \alpha_d) \in (0, \infty)^d$ if for any $h = (h_1, \ldots, h_d)$ with $h_i > 0, i = 1, \ldots, d$, the stochastic process $(\hat{X}_t, t \in T)$ given by

$$\hat{X}_t = h^\alpha X_{\frac{t}{h}} = h_1^{\alpha_1} \ldots h_d^{\alpha_d} X_{\frac{t_1}{h_1}, \ldots, \frac{t_d}{h_d}}, \quad t = (t_1, \ldots, t_d) \in \mathbb{R}_+^d$$

has the same finite-dimensional distributions as the process $(X_t, t \in T)$.

We also the notion of the high-order increment of a d-parameter process X on a rectangle $[s, t] \subset \mathbb{R}^d$, $s = (s_1, \ldots, s_d), t = (t_1, \ldots, t_d)$, with $s < t$. This increment is denoted by $\Delta X_{[s,t]}$ and it is given by

$$\Delta X_{[s,t]} = \sum_{r \in \{0,1\}^d} (-1)^{d - \sum_i r_i} X_{s + \mathbf{r} \cdot (t-s)}. \qquad (3.3)$$

When $d = 1$ one obtains the

$$\Delta X_{[s,t]} = X_t - X_s$$

while for $d = 2$ one gets, if $s = (s_1, s_2), t = (t_1, t_2)$,

$$\Delta X_{[s,t]} = X_{t_1,t_2} - X_{t_1,s_2} - X_{s_1,t_2} + X_{s_1,s_2}.$$

The below definition constitutes a generalization to the multiparameter context of Definition 1.5.

Definition 3.3. Let $T = \mathbb{R}^d$ or $T = \mathbb{R}^d_+$. A process $(X_t, t \in T)$ has stationary increments if for every $s > 0, s \in \mathbb{R}^d$ the stochastic processes

$$(\Delta X_{[0,t]}, t \in T) \text{ and } (\Delta X_{[s,s+t]}, t \in T)$$

have the same finite dimensional distributions.

The anisotropic fractional Brownian sheet is a multi-parameter self-similar process in the sense of Definition 3.2 and its high-order increments are stationary.

Proposition 3.1. *Let* $T = \mathbb{R}^d$ *or* $T = [0, \infty)^d$ *with* $d \geq 1$ *and let* $(W_t^H, t \in \mathbb{R}^d)$ *be a d-parameter fractional Brownian sheet with Hurst parameter* $H = (H_1, \ldots, H_d) \in (0, 1)^d$. *Then* W^H *is a self-similar process with stationary increments in the sense of Definitions 3.2 and 3.3.*

Proof. Fix $h = (h_1, \ldots, h_d)$ with $h_i > 0$ for $i = 1, \ldots, d$. Consider the d-parameter $(X_t, t \in T)$ given by

$$X_t = h_1^{H_1} \ldots h_d^{H_d} W^H_{\frac{t_1}{h_1}, \ldots, \frac{t_d}{h_d}}$$

ifr $t = (t_1, \ldots, t_d) \in T$. It is a centered d-parameter Gaussian process and using (3.1), its covariance function is given by

$$\mathbf{E} X_t X_s = h_1^{2H_1} \ldots h_d^{2H_d} \mathbf{E} W^H_{\frac{t_1}{h_1}, \ldots, \frac{t_d}{h_d}} W^H_{\frac{s_1}{h_1}, \ldots, \frac{s_d}{h_d}}$$

$$= h_1^{2H_1} \ldots h_d^{2H_d} \prod_{i=1}^{d} \left(\frac{1}{2} \left(\left| \frac{t_i}{h_i} \right|^{2H_i} + \left| \frac{s_i}{h_i} \right|^{2H_i} - \left| \frac{t_i - s_i}{h_i} \right|^{2H_i} \right) \right)$$

$$= \prod_{j=1}^{d} \left(\frac{1}{2} (t_j^{2H_j} + s_j^{2H_j} - |t_j - s_j|^{2H_j}) \right)$$

and so X has the same law as W^H. A similar argument (but more tedious to write it down!) shows the stationarity of the increments. $\square$

Remark 3.2. In the one-parameter case $(d = 1)$ the self-similarity and the stationarity of the increments determines the covariance, and thus the probability law, of a Gaussian process, see Theorem 1.2. This is not necessarily the case for the multiparameter Gaussian processes, see [Makogin and Mishura (2019)].

In order to analyze the sample paths of the anisotropic fractional Brownian sheet, we recall the two-parameter version of the Kolmogorov continuity theorem (see e.g. [Ayache *et al.* (2002)]).

Theorem 3.1. *Let* $(X_t, t \in T)$ *be a d-parameter process, vanishing on the axis, with T a compact subset of $\mathbb{R}^d$. Suppose that there exist constants $C, p > 0$ and $\beta_1, \ldots, \beta_d > 1$ such that*

$$\mathbf{E} \left| \Delta X_{[t,t+h]} \right|^p \leq C h_1^{\beta_1} \ldots h_d^{\beta_d}$$

for every $h = (h_1, \ldots, h_d) \in (0, \infty)^d$ and for every $t \in T$ such that $t + h \in T$. Then X admits a continuous modification $\tilde{X}$. Moreover $\tilde{X}$ has Hölder continuous paths of any order $\delta = (\delta_1, \ldots, \delta_d)$ with $\delta_i \in (0, \frac{\beta_i - 1}{p})$ for $i = 1, \ldots, d$ in the following sense: for every $\omega \in \Omega$, there exists $\tilde{C}_\omega > 0$ such that for every $t, t' \in T$

$$\left| \Delta X_{[t,t']}(\omega) \right| \leq C_\omega |t - t'|^\delta$$

where $|t - t'|^\delta = \prod_{j=1}^d |t_j - t'_j|^{\delta_j}$ if $t = (t_1, \ldots, t_d), t' = (t'_1, \ldots, t'_d)$, see convention (3.2).

Using the above criterion we can get the existence of a continuous version of the anisotropic fractional Brownian sheet, as well as the Hölder regularity index for this version.

Proposition 3.2. *Let* $(W_t^H, t \in \mathbb{R}^d)$ *be an anisotropic fractional Brownian sheet with Hurst index $H = (H_1, \ldots, H_d) \in (0, 1)^d$. Then W^H admits a version with Hölder continuous paths of order $\delta = (\delta_1, \ldots, \delta_d)$ such that $\delta_i \in (0, H_i)$ for every $i = 1, \ldots, d$.*

Proof. It can be easily seen that for $t \in \mathbb{R}^d$ and $h = (h_1, \ldots, h_d) \in (0, \infty)^d$, $\Delta W_{[t,t+h]}^H$ is a Gaussian random variable with zero expectation and

$$\mathbf{E} |\Delta W_{[t,t+h]}^H|^2 = \prod_{i=1}^d h_i^{2H_i}$$

so

$$\Delta W_{[t,t+h]}^H \sim N\left(0, \prod_{i=1}^d h_i^{2H_i} \right).$$

Thus for every $p \geq 1$,

$$\mathbf{E} |\Delta W_{[t,t+h]}^H|^2 = \mathbf{E} |Z|^p \left(\prod_{i=1}^d h_i^{2H_i} \right)^{\frac{p}{2}} = \mathbf{E} |Z|^p \prod_{i=1}^d h_i^{pH_i}$$

with $Z \sim N(0,1)$. By choosing $p > \max(H_1, \ldots, H_d)$ large enough, we obtain the conclusion via Theorem 3.1. $\qquad\square$

Remark 3.3. As a particular case of the process introduced in Definition 3.1, let us indicate the anisotropic Wiener sheet $(W_t, t \in \mathbb{R}^d)$ with time interval $\mathbb{R}^d$. It is obtained by taking $H_i = \frac{1}{2}$ for every $i = 1, \ldots, d$ in Definition 3.1. The covariance of the anisotropic Brownian sheet with time interval $\mathbb{R}^d$ can be expressed as

$$\mathbf{E} W_t W_s = \prod_{j=1}^{d} \left(\frac{1}{2}(|t_j| + |s_j| - |t_j - s_j|) \right)$$

$$= \prod_{j=1}^{d} (|t_j| \wedge |s_j|) 1_{(0,\infty)}(t_j s_j) \tag{3.4}$$

if $t = (t_1, \ldots, t_d)$ and $s = (s_1, \ldots, s_d)$.

If one fix one component of the anisotropic fractional Brownian sheet, the process obtained in this way behaves as a "weighted" anisotropic fractional Brownian sheet.

Proposition 3.3. *Let* $T = T_1 \times \ldots \times T_d \subset \mathbb{R}^d$ *and let* $(W_t^H, t \in T \subset \mathbb{R}^d)$ *be an anisotropic fractional Brownian sheet with Hurst index* $H = (H_1, \ldots, H_d) \in (0,1)^d$. *Assume* $t_d \in T_d$ *is fixed. Then the process* $\left(W_{t_1, \ldots, t_{d-1}, t_d}^H, (t_1, \ldots, t_{d-1}) \in T_1 \times \ldots \times T_{d-1} \right)$ *has the same finite dimensional distributions as the process*

$$t_d^{H_d} \left(W_{t_1, \ldots, t_{d-1}}^{(H_1, \ldots, H_{d-1})}, (t_1, \ldots, t_{d-1}) \in T_1 \times \ldots \times T_{d-1} \right)$$

where $\left(W_{t_1, \ldots, t_{d-1}}^{(H_1, \ldots, H_{d-1})}, (t_1, \ldots, t_{d-1}) \in T_1 \times \ldots \times T_{d-1} \right)$ *is a* $(d-1)$-*parameter anisotropic fractional Brownian motion with Hurst index* $(H_1, \ldots, H_{d-1}) \in (0,1)^{d-1}$.

Proof. It is easy to see that for $t_d \in T_d$, by (3.1),

$$\mathbf{E} W_{t_1, \ldots, t_{d-1}, t_d}^H W_{s_1, \ldots, s_{d-1}, t_d}^H = t_d^{2H_d} \prod_{j=1}^{d-1} R_{H_j}(t_j, s_j)$$

$$= \mathbf{E} \left(t_d^{H_d} W_{t_1, \ldots, t_{d-1}}^{(H_1, \ldots, H_{d-1})} t_d^{2H_d} W_{s_1, \ldots, s_{d-1}}^{(H_1, \ldots, H_{d-1})} \right)$$

for every $t_i, s_i \in T_i$ for $i = 1, \ldots, d-1$. $\qquad \square$

3.2 The space-time white noise

The space-time white noise can be viewed as a particular case of the Brownian sheet introduced in the previous section, see Remark 3.3. Since it plays an important role, let us detail its definition and properties. Let us denote by $\mathcal{B}_b(\mathbb{R}^d)$ the set of bounded Borel subsets of $\mathbb{R}^d$.

Definition 3.4. Let $T = \mathbb{R}_+ \times \mathcal{B}_b(\mathbb{R}^d)$. The space-time white noise is a centered Gaussian random field $(W(x), x \in T)$ with covariance function

$$\mathbf{E}W(x)W(y) = (t \wedge s)\lambda(A \cap B) \text{ if } x = (t, A), y = (s, B) \qquad (3.5)$$

with $t, s \geq 0$ and $A, B \in \mathcal{B}_b(\mathbb{R}^d)$. We denoted by λ the Lebesgue measure on $\mathbb{R}^d$.

The existence of the space-time white noise is ensured by the following result.

Proposition 3.4. *Let* $T = \mathbb{R}_+ \times \mathcal{B}_b(\mathbb{R}^d)$ *and consider the function* $R : T^2 \to \mathbb{R}$,

$$R((t, A), (s, B)) = (t \wedge s)\lambda(A \cap B) \text{ for } t, s \geq 0, A, B \in \mathcal{B}_b(\mathbb{R}^d).$$

Then R *is summetric and positive definite.*

Proof. R is obviously symmetric. Let $N \geq 1$, $c_1, \ldots, c_n \in \mathbb{R}$ and $(t_i, A_i) \in T$ for $i = 1, \ldots, N$. We can write

$$\sum_{i,j=1}^{N} c_i c_j R\left((t_i, A_i), (t_j, A_j)\right)$$

$$= \sum_{i,j=1}^{N} c_i c_j (t_i \wedge t_j)\lambda(A_i \cap A_j)$$

$$= \sum_{i,j=1}^{N} c_i c_j \int_{\mathbb{R}_+ \times \mathbb{R}^d} 1_{[0,t_i] \times A_i}(s, y) 1_{[0,t_j] \times A_j}(s, y) ds dy$$

$$= \int_{\mathbb{R}_+ \times \mathbb{R}^d} \left(\sum_{i=1}^{N} c_i 1_{[0,t_i] \times A_i}(s, y)\right)^2 ds dy \geq 0.$$

$\square$

Let us discuss the behavior of the space-time white noise with respect to each of its two variables.

Remark 3.4.

- Let $T = \mathbb{R}_+ \times \mathcal{B}_b(\mathbb{R}^d)$ and let $(W(x), x \in T)$ be a space-time white noise. Fix $A \in \mathbb{B}_b(\mathbb{R}^d)$. Then the process $(W(t, A), t \geq 0)$ has the same law as

$$(\sqrt{\lambda(A)}B_t, t \geq 0)$$

where B is a standard Brownian motion. This comes from the relation

$$\mathbf{E}W(t, A)W(s, A) = (t \wedge s)\lambda(A)$$

for every $s, t \geq 0$. Consequently, the (one-parameter) Gaussian process $(W(t, A), t \geq 0)$ is $\frac{1}{2}$-self-similar. Moreover, its increments are independent i.e.

$$W(t, A) - W(s, A) \text{ and } W(u, A) - W(t, A)$$

are independent random variables if $s < t < u$.

- Let $t \geq 0$ be fixed. Then $(W(t, A), A \in \mathcal{B}_b(\mathbb{R}^d))$ is a Gaussian random field with zero expectation and covariance

$$\mathbf{E}W(t, A)W(t, B) = t\lambda(A \cap B)$$

for every $A, B \in \mathcal{B}_b(\mathbb{R}^d)$. This implies that $(W(t, A), A \in \mathcal{B}_b(\mathbb{R}^d))$ "behaves" as

$$\left(\sqrt{t}W(x), x \in \mathbb{R}\right)$$

where $(W(x), x \in \mathbb{R})$ is a d-parameter anisotropic Brownian motion. We refer to the next paragraph for the details.

- The above properties motivate the name of this process: it is called a space-time white noise because it behaves as Wiener process with respect to its time variable and as a Wiener sheet with respect to the spatial variable.

3.2.1 *The space-time white noise and the anisotropic Brownian sheet*

If $(W(t, A), t \geq 0, A \in \mathcal{B}_b(\mathbb{R}^d))$ is a space-time white noise, we can associate to it an anisotropic Brownian sheet. Let us consider first $d = 1$. Define the centered Gaussian random field $(M(t, x), t \geq 0, x \in \mathbb{R})$ as follows:

$$M(t, x) = \begin{cases} W(t, [0, x]) & \text{if } t \geq 0, x \geq 0 \\ W(t, [x, 0]) & \text{if } t \geq 0, x < 0. \end{cases} \tag{3.6}$$

Then the random field $(M(t,x), t \geq 0, x \in \mathbb{R})$ defined by (3.6) is a two-parameter anisotropic Brownian motion with time interval $T = \mathbb{R}_+ \times \mathbb{R}$. Indeed, for every $s, t \geq 0$ and $x, y \in \mathbb{R}$ we have

$$
\mathbf{E}M(t,x)M(s,y) = \begin{cases} (t \wedge s)(x \wedge y) \text{ if } x, y \geq 0 \\ 0 \text{ if } x < 0 \text{ or } y < 0 \\ -(x \vee y)(t \wedge s) = ((-x) \wedge (-y))(t \wedge s) \text{ if } x, y < 0 \end{cases}
$$

so

$$
\mathbf{E}M(t,x)M(s,y) = (t \wedge s)(|x| \wedge |y|)1_{(0,\infty)}(xy).
$$

We retrieve the covariance of the anisotropic Brownian sheet (compare with (3.4)). The construction can be generalize to any dimension $d \geq 1$ in the following way: for $x = (x_1, \ldots, x_d) \in \mathbb{R}^d$, set

$$
M(t,x) = W\left(t, \prod_{j=1}^{d} \left([0, x_j]1_{x_j \geq 0} + [x_j, 0]1_{x_j < 0}\right)\right).
$$

Then the random field $(M(t,x), t \geq 0, x \in \mathbb{R}^d)$ is a centered Gaussian process with covariance given by

$\mathbf{E}M(t,x)M(s,y)$

$$
= (t \wedge s)\lambda \left[\prod_{j=1}^{d} \left([0, x_j]1_{x_j \geq 0} + [x_j, 0]1_{x_j < 0}\right) \bigcap \prod_{j=1}^{d} \left([0, y_j]1_{y_j \geq 0} + [y_j, 0]1_{y_j < 0}\right)\right]
$$

$$
= \begin{cases} (t \wedge s) \prod_{j=1}^{d}(|x_j| \wedge |y_j|) \text{ if } x_j y_j \geq 0 \text{ for every } j = 1, \ldots, d \\ 0, \text{ if there exists } j = 1, \ldots, d \text{ such that } x_j y_j < 0. \end{cases}
$$

Therefore its covariance coincides with (3.4), so it a $(d+1)$-parameter anisotropic Wiener process.

Conversely, given an anisotropic Brownian sheet $(M(t,x), t \geq 0, x \in \mathbb{R}^d)$, we set for any rectangle $A \in \mathcal{B}_b(\mathbb{R}^d)$ of the form $A = [a_1, b_1] \times \ldots [a_d, b_d]$ with $a_i < b_i, i = 1, \ldots, d$

$$
W(t, A) = \Delta M(t, \cdot)_{[a,b]}. \tag{3.7}
$$

Here $\Delta M(t, \cdot)_{[a,b]}$ denotes the d-dimensional high-order increment of the process $(M(t,x), x \in \mathbb{R}^d)$ over the rectangle $[a,b]$, see (3.3). For instance, if $d = 1$

$$
\Delta M(t, \cdot)_{[a_1, b_1]} = M(t, b_1) - M(t, a_1)
$$

and for $d = 2$,

$$\Delta M(t, \cdot)_{[a_1, b_1] \times [a_2, b_2]}$$
$$= M(t, (b_1, b_2)) - M(t, (a_1, b_2)) - M(t, (a_2, b_1)) + M(t, (a_1, a_2)).$$

The definition (3.7) can be extended to any $A \in \mathcal{B}_b(\mathbb{R}^d))$ and the Gaussian random field $(W(t, A), t \geq 0, A \in \mathcal{B}_b(\mathbb{R}^d))$ obtained in this way becomes a space-time white noise. For instance, if $d = 1$ and $0 \leq a < b, 0 \leq c < d$

$$\mathbf{E}W(t, [a, b])W(s, [c, d])$$
$$= \mathbf{E}\left((M(t, b) - M(t, a))(M(s, d) - M(s, c))\right)$$
$$= (t \wedge s)\left((b \wedge d) + (a \wedge c) - (b \wedge c) - (a \wedge d)\right)$$
$$= (t \wedge s)\lambda([a, b] \cap [c, d]).$$

3.3 Isotropic fractional Brownian sheet

Let us now define another multiparameter stochastic process which generalizes the one-dimensional fractional Brownian motion. Below $\| \cdot \|$ denotes the Euclidean norm in $\mathbb{R}^d$.

Definition 3.5. The d-parameter isotropic fractional Brownian motion with Hurst index $H \in (0, 1)$ is a centered Gaussian process $(B^H(x), x \in \mathbb{R}^d)$ with covariance function

$$\mathbf{E}B^H(x)B^H(y) = \frac{1}{2}\left(\|x\|^{2H} + \|y\|^{2H} - \|x - y\|^{2H}\right) \qquad (3.8)$$

for every $x, y \in \mathbb{R}^d$.

Remark 3.5.

- Proposition 2.1 ensures the fact the the isotropic fBm exists.
- By taking $H = \frac{1}{2}$ in (3.8), we obtain the isotropic Wiener process, i.e. a centered Gaussian process $(W(x), x \in \mathbb{R}^d)$ with covariance

$$\mathbf{E}W(x)W(y) = \frac{1}{2}\left(\|x\| + \|y\| - \|x - y\|\right), \qquad (3.9)$$

for every $x, y \in \mathbb{R}^d$.

- We can also define the isotropic fractional Brownian sheet for the time indexed by a set $T \subset \mathbb{R}^d$.

An important property, which makes this Gaussian sheet different from the anisotropic fractional Brownian motion is that for every $s, t \in \mathbb{R}^d$

$$\mathbf{E}\left(B^H(\mathbf{x}) - B^H(\mathbf{y})\right)^2 = \|\mathbf{x} - \mathbf{y}\|^{2H}$$

which implies, due to the Gaussianity, that for every $n \geq 1$

$$\mathbf{E} \left(B^H(\mathbf{x}) - B^H(\mathbf{y}) \right)^n = \mathbf{E}|Z|^n \|\mathbf{x} - \mathbf{y}\|^{nH}, \tag{3.10}$$

where Z is a standard normal random variable. From (3.10) one can deduce, the existence of a continuous version for B^H via the following multi-dimension Kolmogorov continuity criterion, which constitutes an alternative to Theorem 3.1 (see e.g. [Khosnevisan (2002)]).

Theorem 3.2. *Let $(X_t, t \in T)$ be a d-parameter process indexed by $T = [a_1, b_1] \times \ldots \times [a_d, b_d]$. Assume that there exist $C, p > 0$ and $\beta > d$ such that*

$$\mathbf{E} |X_t - X_s|^p \leq C \|t - s\|^{\beta}$$

uniformly in $s, t \in T$. Then X admits a continuous version. Moreover, for any $0 < \theta < \frac{\beta - d}{p}$,

$$\mathbf{E} \left| \sup_{s \neq t} \frac{|X_t - X_s|}{\|t - s\|^{\theta}} \right|^p < \infty. \tag{3.11}$$

Above $\|x\|$ may be anyone of the following norms on $\mathbb{R}^d$, if $x = (x_1, \ldots, x_d)$:

$$\|x\| = \max \left(|x_1|, \ldots, |x_d| \right)$$
$$\|x\| = \left(|x_1|^p + \ldots |x_d|^p \right)^{\frac{1}{p}}, \ for \ p \geq 1$$
$$\|x\| = |x_1|^p + \ldots |x_d|^p, \ for \ 0 < p < 1.$$

Proposition 3.5. *Let $(B^H(x), x \in \mathbb{R}^d)$ be a d-parameter isotropic fBm with $H \in (0, 1)$. Then B^H admits a continuous version and it satisfies (3.11) for $0 < \theta < H$.*

Proof. If follows from (3.10) and Theorem 3.2. $\qquad\qquad\qquad\square$

The isotropic fBm satisfies a different notion of self-similarity and stationary increments.

Proposition 3.6. *Let $T = \mathbb{R}^d$ or $T = \mathbb{R}_+^d$. Let $(B^H(x), x \in \mathbb{R}^d)$ be a d-parameter isotropic fBm with $H \in (0, 1)$. Then for every $h = (h_1, \ldots, h_d) \in \mathbb{R}_+^d$, the d-parameter process*

$$\left(B^H(x + h) - B^H(h), x \in T \right)$$

has the same law as B^H.

Proof. Let $h = (h_1, \ldots, h_d) \in \mathbb{R}_+^d$. Then $\left(B^H(x+h) - B^H(h), x \in T\right)$ is a centered Gaussian process. By (3.8),

$$\mathbf{E}(B^H(x+h) - B^H(h))(B^H(y+h) - B^H(h))$$

$$= \frac{1}{2} \left[\|x+h\|^{2H} + \|y+h\|^{2H} - \|x-y\|^{2H} - \left(\|x+h\|^{2H} + \|h\|^{2H} - \|x\|^{2H} \right) \right.$$

$$\left. - \left(\|y+h\|^{2H} + \|h\|^{2H} - \|y\|^{2H} \right) + 2\|h\|^{2H} \right]$$

$$= \frac{1}{2} \left(\|x\|^{2H} + \|y\|^{2H} - \|x-y\|^{2H} \right)$$

and this shows that it has the same finite dimensional distributions as B^H. $\qquad\square$

We have the following scaling property.

Proposition 3.7. *Let* $T = \mathbb{R}^d$ *or* $T = \mathbb{R}_+^d$. *Let* $(B^H(x), x \in \mathbb{R}^d)$ *be a d-parameter isotropic fBm with* $H \in (0, 1)$. *Then for every* $a > 0$, *the process* $(B^H(ax), x \in T)$ *has the same law as* $(a^H B^H(x), x \in T)$.

Proof. Both processes from the statement are centered Gaussian processes and by (3.8), their covariance is

$$\frac{1}{2} a^{2H} \left(\|x\|^{2H} + \|y\|^{2H} - \|x-y\|^{2H} \right)$$

for every $a > 0, x, y \in T$. $\qquad\square$

3.4 The white-colored Gaussian noise

As its name indicates, the white-colored Gaussian noise is a multiparameter Gaussian process which is "white" in time (i.e. it behaves as a Wiener process with respect to its time variable) and it has a "colored" spatial covariance, i.e. its spatial increments are correlated. It is often chosen to model the random perturbation which drives various stochastic partial differential equations.

The result below will justify the existence of the white-colored Gaussian noise.

Proposition 3.8. *Let* $T = \mathbb{R}_+ \times \mathcal{B}_b(\mathbb{R}^d)$. *Define the function* $C : T^2 \to \mathbb{R}$ *given by*

$$C\left((t, A), (s, B)\right) = (t \wedge s) \int_A \int_B \|x-y\|^{-(d-\alpha)} dx dy, \quad t, s \geq 0, A, B \in \mathcal{B}_b(\mathbb{R}^d)$$

with $0 < \alpha < d$. *Then* C *is a covariance function.*

Proof. For every $\lambda_1, \ldots, \lambda_N \in \mathbb{R}$ and $(t_i, A_i) \in T$ for $i = 1, \ldots, N$ we have

$$\sum_{i,j=1}^{N} \lambda_i \lambda_j C\left((t_i, A_i), (t_j, A_j)\right)$$

$$= \sum_{i,j=1}^{N} \lambda_i \lambda_j (t_i \wedge t_j) \int_{A_i} \int_{A_j} \|x - y\|^{-(d-\alpha)} dx dy$$

$$= \sum_{i,j=1}^{N} \lambda_i \lambda_j (t_i \wedge t_j) \int_{\mathbb{R}^d} \int_{\mathbb{R}^d} dx dy 1_{A_i}(x) 1_{A_j}(y) \|x - y\|^{-(d-\alpha)}.$$

We denotes by $\mathcal{F}f$ the Fourier transform of f (see (5.9)). By using the Parseval's relation (6.3), with $c_\alpha > 0$,

$$\sum_{i,j=1}^{N} \lambda_i \lambda_j C\left((t_i, A_i), (t_j, A_j)\right)$$

$$= c_\alpha \sum_{i,j=1}^{N} \lambda_i \lambda_j (t_i \wedge t_j) \int_{\mathbb{R}^d} d\xi \mathcal{F} 1_{A_i}(\cdot)(\xi) \overline{\mathcal{F} 1_{A_j}(\cdot)}(\xi) \|\xi\|^{-\alpha}$$

$$= c_\alpha \sum_{i,j=1}^{N} \lambda_i \lambda_j \int_{\mathbb{R}_+} ds \int_{\mathbb{R}^d} d\xi 1_{[0,t_i]}(s) 1_{[0,t_j]}(s) \mathcal{F} 1_{A_i}(\cdot)(\xi) \overline{\mathcal{F} 1_{A_j}(\cdot)}(\xi) \|\xi\|^{-\alpha}$$

$$= c_\alpha \int_{\mathbb{R}_+} ds \int_{\mathbb{R}^d} d\xi \|\xi\|^{-\alpha} \left| \sum_{i=1}^{N} \lambda_i 1_{[0,t_i]}(s) \mathcal{F} 1_{A_i}(\cdot)(\xi) \right|^2 \geq 0.$$

$\square$

Let us now define the white-colored Gaussian noise.

Definition 3.6. Let $T = \mathbb{R}_+ \times \mathcal{B}_b(\mathbb{R}^d)$. The white-colored noise (with spatial covariance given by the Riesz kernel) is defined as a zero mean Gaussian field $(W^\gamma(x), x \in T)$ with covariance

$$\mathbf{E} W^\gamma(t, A) W^\gamma(s, B) = (t \wedge s) \int_A \int_B f(x - y) dx dy \qquad (3.12)$$

for every $s, t \geq 0$, $A, B \in \mathcal{B}_b(\mathbb{R}^d)$, where f is the Riesz kernel of order $\gamma \in (0, d)$ defined by

$$f(x) = R_\gamma(x) := g_{\gamma,d} \|x\|^{-d+\gamma}, \quad 0 < \alpha < d, \qquad (3.13)$$

where $g_{\gamma,d} = 2^{d-\gamma} \pi^{d/2} \Gamma((d - \gamma)/2) / \Gamma(\gamma/2)$.

By Proposition 3.8, we deduce the existence of the white-colored Gaussian noise. This process has the following behavior if one of its components is fixed.

Remark 3.6.

(1) Let $(W^\gamma(x), x \in T)$ be a white-colored Gaussian noise. Then for every fixed $A \in \mathcal{B}_b(\mathbb{R}^d)$, the stochastic process $(W^\gamma(t, A), t \geq 0)$ coincides in distribution with

$$\left(\sqrt{C_A} W_t, t \geq 0\right)$$

where $(W_t, t \geq 0)$ is a standard Wiener process and $C_A = \int_A \int_A f(x - y) dx dy < \infty$.

(2) Let $t > 0$ be fixed. Then $(W^\gamma(t, A), A \in \mathcal{B}_b(\mathbb{R}^d))$ is a centered Gaussian field with covariance

$$\mathbf{E} W^\gamma(t, A) W^\gamma(t, B) = t \int_A \int_B f(x - y) dx dy$$

for every $A, B \in \mathcal{B}_b(\mathbb{R}^d)$. For $d = 1$, we show below in Section 3.4.1 that is behaves, modulo a constant, as a fractional Brownian motion.

3.4.1 *The white-colored Gaussian noise and the anisotropic fractional Brownian sheet*

Let us discuss the relation between the anisotropic fractional Brownian sheet and the white-colored Gaussian noise. Assume $d = 1$ and let $(W^\gamma(t, A), t \geq 0, A \in \mathbb{B}_b(\mathbb{R}))$ be a centered Gaussian family with covariance (3.12), with $\gamma \in (0, 1)$. Define the random field $(M^\gamma(t, x), t \geq 0, x \in \mathbb{R})$ as follows:

$$M^\gamma(t, x) = \begin{cases} W^\gamma(t, [0, x]) \text{ if } t \geq 0, x \geq 0 \\ -W^\gamma(t, [x, 0]) \text{ if } t \geq 0, x < 0. \end{cases}$$

We use below the relation, for $\gamma \in (0, 1)$,

$$\int_{x_1}^x \int_{y_1}^y da db |a - b|^{\gamma-1} \tag{3.14}$$
$$= (\gamma(\gamma + 1))^{-1} \left(|x - y_1|^{\gamma+1} + |x_1 - y|^{\gamma+1} - |x - y|^{2\gamma} - |x_1 - y_1|^{\gamma+1}\right)$$

Then $(M^\gamma(t, x), t \geq 0, x \in \mathbb{R})$ is a centered Gaussian random field with the following covariance:

(1) If $t, s \geq 0, x, y \geq 0$, then by (3.14), with $g_{\gamma,1}$ from (3.13),

$$\mathbf{E}M^\gamma(t,x)M^\gamma(s,y)$$
$$= \mathbf{E}W^\gamma(t,[0,x])W^\gamma(s,[0,y])$$
$$= g_{\gamma,1}(t \wedge s) \int_0^x \int_0^y |z - z'|^{\gamma-1} dz dz'$$
$$= g_{\gamma,1}(t \wedge s)(\gamma(\gamma+1))^{-1} \left(x^{1+\gamma} + y^{1+\gamma} - |x-y|^{1+\gamma} \right)$$

where the last equality follows by (3.14).

(2) If $s, t \geq 0, x \geq 0, y < 0$, then

$$\mathbf{E}M^\gamma(t,x)M^\gamma(s,y)$$
$$= \mathbf{E}W^\gamma(t,[0,x])W^\gamma(s,[y,0])$$
$$= g_{\gamma,1}(t \wedge s) \int_0^x \int_y^0 |z - z'|^{\gamma-1} dz dz'$$
$$= g_{\gamma,1}(t \wedge s)(\gamma(\gamma+1))^{-1} \left((-x)^{1+\gamma} + (-y)^{1+\gamma} - |x-y|^{1+\gamma} \right)$$
$$= g_{\gamma,1}(t \wedge s)(\gamma(\gamma+1))^{-1} \left(|x|^{1+\gamma} + |y|^{1+\gamma} - |x-y|^{1+\gamma} \right)$$

The same holds for $x < 0, y \geq 0$.

(3) If $t, s \geq 0, x, y < 0$ then

$$\mathbf{E}M^\gamma(t,x)M^\gamma(s,y)$$
$$= \mathbf{E}W^\gamma(t,[0,x])W^\gamma(s,[y,0])$$
$$= g_{\gamma,1}(t \wedge s) \int_x^0 \int_y^0 |z - z'|^{\gamma-1} dz dz'$$
$$= g_{\gamma,1}(t \wedge s)(\gamma(\gamma+1))^{-1} \left(-x^{1+\gamma} - y^{1+\gamma} - |x-y|^{1+\gamma} \right)$$
$$= g_{\gamma,1}(t \wedge s)(\gamma(\gamma+1))^{-1} \left(|x|^{1+\gamma} + |y|^{1+\gamma} - |x-y|^{1+\gamma} \right)$$

In conclusion, $(M^\gamma(t,x), t \geq 0, x \in \mathbb{R})$ is, modulo a constant, an anisotropic fractional Brownian sheet with Hurst index $\left(\frac{1}{2}, \frac{1+\gamma}{2} \right)$, i.e. it is a centered Gaussian field, indexed on $T = \mathbb{R}_+ \times \mathbb{R}$, with covariance

$$\mathbf{E}M^\gamma(t,x)M^\gamma(s,y) = C_0(t \wedge s)R_H(x,y) \qquad (3.15)$$

with $C_0 = 2g_{\gamma,1}(\gamma(\gamma+1))^{-1}$, $H = \left(\frac{1}{2}, \frac{1+\gamma}{2} \right)$ and R_H defined by (3.1).

Conversely, let $(M^\gamma(t,x), t \geq 0, x \in \mathbb{R})$ be a centered Gaussian field with covariance (3.15). If $A = [a_1, a_2]$ with $a_1 < a_2$, we set

$$W^\gamma(t,A) = M^\gamma(t,a_2) - M^\gamma(t,a_1) \text{ for } t \geq 0.$$

Then above definition can to naturally extended to any $A \in \mathcal{B}_b(\mathbb{R})$ and the random field $(W^\gamma(t,A), t \geq 0, A \in \mathcal{B}_b(\mathbb{R}))$ is a white-colored Gaussian noise.

3.4.2 Other spatial covariance kernels

There are other ways to define the spatial "color" for a Gaussian noise, by replacing the Riesz kernel (3.13) by others kernels. Here are some well-known examples. Everywhere below, $x \in \mathbb{R}^d$.

- The Bessel kernel of order α

$$f(x) = B_\alpha(x) := \gamma'_\alpha \int_0^\infty w^{(\alpha-d)/2-1} e^{-w} e^{-\|x\|^2/(4w)} dw, \quad \alpha > 0,$$

where $\gamma'_\alpha = (4\pi)^{\alpha/2} \Gamma(\alpha/2)$.

- The Poisson kernel of order α given by

$$f(x) = P_\alpha(x) := \gamma'''_{\alpha,d}(\|x\|^2 + \alpha^2)^{-(d+1)/2}, \quad \alpha > 0,$$

where $\gamma'''_{\alpha,d} = \pi^{-(d+1)/2} \Gamma((d+1)/2)\alpha$.

- The heat kernel of order α

$$f(x) = G_\alpha(x) := \gamma''_{\alpha,d} e^{-\|x\|^2/(4\alpha)}, \quad \alpha > 0,$$

where $\gamma''_{\alpha,d} = (4\pi\alpha)^{-d/2}$.

One can show that with all of these kernels instead on the Riesz kernel in (3.12), we have a well-defined covariance function. On the other hand, the particular choice of the Riesz kernel for the spatial covariance of the random noise of the stochastic heat equation leads to a solution whose probability law is related to the fractional Brownian motion. These aspects will be discussed in the next chapters.

Chapter 4

Isonormal processes and Wiener integral

The isonormal processes are Gaussian processes indexed by a Hilbert space with a particular form of their covariance function. They are the basis of the construction of the Wiener chaos and of the Malliavin calculus (see e.g. [Nualart (2006)]). We will see that one can associate an isonormal process to any of the Gaussian processes and sheets presented above in Chapter 3. This link is equivalent to the construction of the Wiener integral with respect to these one-parameter and multiparameter Gaussian processes.

4.1 Definition and general properties

Let H be a real and separable Hilbert space. Denote by $\langle \cdot, \cdot \rangle_H$ and $\| \cdot \|_H$ the scalar product and the norm in H, respectively.

Definition 4.1. An isonormal process is a centered Gaussian family $(W(h), h \in H)$ such that

$$\mathbf{E}W(h)W(g) = \langle h, g \rangle_H \text{ for every } h, g \in H. \tag{4.1}$$

Let us first notice that the right-hand side of (4.1) defines a covariance function. Consider the function

$$C : H^2 \to \mathbb{R}, \quad C(h, g) = \langle h, g \rangle_H.$$

Then, for every $N \geq 1$ and for $\lambda_1, \ldots, \lambda_N \in \mathbb{R}$, $h_1, \ldots, h_N \in H$ arbitrary chosen,

$$\sum_{i,j=1}^{N} \lambda_i \lambda_j C(h_i, h_j) = \sum_{i,j=1}^{N} \lambda_i \lambda_j \langle h_i, h_j \rangle_H$$

$$= \left\| \sum_{i=1}^{N} \lambda_i h_i \right\|_H^2 \geq 0.$$

From the Definition 4.1, we know that for every $\alpha_1, \ldots, \alpha_n \in \mathbb{R}$ and for every $h_1, \ldots, h_n \in H$, the random variable

$$\alpha_1 W(h_1) + \ldots + \alpha_n W(h_n)$$

is Gaussian. Also, for every $h \in H$ we have from (4.1) that

$$\mathbf{E}W(h) = 0 \text{ and } \mathbf{E}W(h)^2 = \|h\|_H^2,$$

so

$$W(h) \sim N(0, \|h\|^2). \tag{4.2}$$

Proposition 4.1. *The application $h \to W(h)$ from H onto $L^2(\Omega)$ is linear.*

Proof. One can show that for every $\alpha \in \mathbb{R}$ and $h, g \in H$ we have

$$W(\alpha h + g) = \alpha W(h) + W(g) \text{ almost surely.}$$

To this end, we show that

$$\mathbf{E}\left|W(\alpha h + g) - (\alpha W(h) + W(g))\right|^2 = 0.$$

Indeed, by (4.1)

$$
\begin{aligned}
&\mathbf{E}\left|W(\alpha h + g) - (\alpha W(h) + W(g))\right|^2 \\
&= \|\alpha h + g\|_H^2 + \alpha^2 \|h\|^2 + 2\alpha\langle h, g\rangle_H + \|g\|_H^2 \\
&\quad -2\left(\langle \alpha h + g, \alpha h\rangle_H + \langle \alpha h + g, g\rangle_H\right) \\
&= 2\alpha^2 \|h\|_H^2 + 4\alpha\langle h, g\rangle_H + \|g\|_H^2 \\
&\quad -2\left(\alpha^2 \|h\|_H^2 + 2\langle h, g\rangle_H + \|g\|_H^2\right) = 0
\end{aligned}
$$

where we used the bilinearity of the scalar product in H. $\qquad\square$

Let us make some additional comments concerning the isonormal processes.

Remark 4.1.

- We assumed that the family $(W(h), h \in H)$ is a Gaussian family. From linearity, we can assume in the definition of the isonomal process that only the random variable $W(h)$ is Gaussian, for every $h \in H$.
- The mapping $h \to W(h)$ from H on $L^2(\Omega)$ is an isometry, as a direct consequence of (4.1).

- For every $h, g \in H$, the random variables $W(h)$ and $W(g)$ are independent if and only $\langle h, g \rangle_H = 0$. This comes from the fact $(W(h), W(g))$ is a Gaussian random vector, so its components are independent if and only they are uncorrelated, i.e.

$$\mathbf{E} W(h) W(g) = \langle h, g \rangle_H = 0,$$

see (1.2).

Let us give a first example of an isonormal process.

Example 4.1. Take H a real and separable Hilbert space. Let $(e_j)_{j \geq 1}$ be an orthonormal basis in H and let $(X_j)_{j \geq 1}$ be a sequence of independent, $N(0, 1)$-distributed random variables. Set

$$W(h) = \sum_{j \geq 1} \langle h, e_j \rangle_H X_j \text{ for any } h \in H.$$

Then $(W(h), h \in H)$ is an isonormal process. It is clear that this is a Gaussian family of centered random variables and for every $h, g \in H$, by the independence of X_js,

$$\begin{aligned}
\mathbf{E} W(h) W(g) &= \sum_{j,k \geq 1} \langle h, e_j \rangle_H \langle g, e_k \rangle_H \mathbf{E} X_j X_k \\
&= \sum_{j \geq 1} \langle h, e_j \rangle_H \langle g, e_j \rangle_H = \langle h, g \rangle_H.
\end{aligned}$$

We discuss the particular case when H is an Hilbert space of L^2-type. Consider an isonormal process $(W(h), h \in H)$ where $H = L^2(T, \mathcal{B}, \mu)$ with $\mathcal{B}$ a sigma-algebra included in $\mathcal{B}(T)$ and μ a sigma-finite measure on $\mathcal{B}$ without atoms. Let

$$W(A) := W(1_A) \text{ for } A \in \mathcal{B} \text{ with } \mu(A) < \infty.$$

Then the family $(W(A), A \in \mathcal{B}, \mu(A) < \infty)$ satisfies the following properties:

(1) For every $A \in \mathcal{B}, \mu(A) < \infty$, $W(A) \sim N(0, \mu(A))$, by (4.2).
(2) For every $A, B \in \mathcal{B}, \mu(A), \mu(B) < \infty$ we have

$$\mathbf{E} W(A) W(B) = \langle 1_A, 1_B \rangle_{H=L^2(T,\mathcal{B},\mu)} = \mu(A \cap B).$$

Thus the random variables $W(A)$ and $W(B)$ are independent if and only if $A \cap B = \emptyset$, due to (1.2).

(3) Assume $A = \cup_{j\geq 1} A_j$ where for every $j \geq 1$, $A_j, A \in \mathcal{B}$ with $\mu(A_j), \mu(A) < \infty$ and $(A_j)_{j\geq 1}$ are mutually disjoint. Then

$$W(A) = \sum_{j\geq 1} W(A_j) \text{ in } L^2(\Omega).$$

First notice that

$$\mathbf{E} \left| \sum_{j\geq 1} W(A_j) \right|^2 = \sum_{j\geq 1} \mu(A_j) = \mu(A) < \infty$$

and

$$\mathbf{E} \left| W(A) - \sum_{j\geq 1} W(A_j) \right|^2 = 2\mu(A) - 2\sum_{j\geq 1} \mathbf{E} W(A)W(A_j))$$

$$= 2\mu(A) - 2\sum_{j\geq 1} \mu(A \cap A_j)$$

$$= 2\mu(A) - 2\mu(A) = 0.$$

Therefore, to any isonormal process $(W(h), h \in H = L^2(T, \mathcal{B}, \mu))$ one can associate a Gaussian family $(W(A), A \in \mathcal{B}, \mu(A) < \infty)$ with the properties 1.-3. above. Conversely, consider a Gaussian family $(M(A), A \in \mathcal{B}, \mu(A) < \infty)$ such that $\mathbf{E} M(A) = 0$ for any $A \in \mathcal{B}, \mu(A) < \infty$ and

$$\mathbf{E} M(A)M(B) = \mu(A \cap B) \text{ if } A, B \in \mathcal{B}, \mu(A), \mu(B) < \infty.$$

Then we can construct an isonormal process $(W(h), h \in H)$ such that $W(1_A) = M(A)$ for every $A \in \mathcal{B}$. Let $\mathcal{E}$ be the set of simple functions of the form

$$h = \sum_{i=1}^{N} \lambda_i 1_{A_i} \tag{4.3}$$

with $\lambda_i \in \mathbb{R}, A_i \in \mathcal{B}, \mu(A_i) < \infty$ for every $i \geq 1$ and the sets A_i are disjoints. For h of the form (4.3), we put

$$W(h) = \sum_{i=1}^{N} \lambda_i M(A_i).$$

The set $\mathcal{E}$ is dense in H. For every $h \in H$ take a sequence $(h_n)_{n\geq 1}$ in $\mathcal{E}$ of the form (4.3) and set

$$W(h) = \lim_{n\to\infty} M(h_n) \text{ in } L^2(\Omega).$$

The above limit exists because $(M(h_n))_{n\geq 1}$ is a Cauchy sequence in $L^2(\Omega)$. Indeed, for every $n, m \geq 1$,

$$\mathbf{E}(M(h_n) - M(h_m))^2 = \mathbf{E}\left(M(h_n - h_m)^2\right) = \|h_n - h_m\|_H^2$$
$$\to 0 \text{ as } m, n \to \infty.$$

One can also show that the limit does not depend on the chosen sequence in $\mathcal{E}$. The family $(W(h), h \in H)$ becomes an isonormal process.

4.2 The Brownian motion as an isonormal process

Let $T \subset \mathbb{R}$ and let $(W_t, t \in T)$ be a Wiener process. For $A \in \mathcal{B}_b(T)$ (the set of Borel sets in T with finite Lebesgue measure), we define the random variable $W(A)$ as follows: if $A = [a, b]$ with $a, b \in T, a < b$, then

$$W(A) = W_b - W_a.$$

Then $(W(A), A \in \mathcal{B}_b(T))$ constitutes a Gaussian family of centered random variables such that

$$\mathbf{E}W(A)W(B) = \lambda(A \cap B), \quad A, B \in \mathcal{B}_b(T)$$

where λ stands for the Lebesgue measure on $\mathbb{R}$. As in the previous section, we can associate to this family an isonormal process $(W(h), h \in H = L^2(T, \mathcal{B}(T), \lambda))$.

For $h \in H = L^2(T, \mathcal{B}(T), \lambda)$, we use the notation

$$W(h) = \int_T h(s)dB_s.$$

This will be called the Wiener integral of h with respect to the Brownian motion B and it satisfies the isometry

$$\mathbf{E}W(h)W(g) = \langle h, g \rangle_{L^2(T)} = \int_T h(s)g(s)ds \text{ if } h, g \in L^2(T). \tag{4.4}$$

Next, we will present some important examples of Wiener integrals. Let us prove that the fBm can be expressed as a Wiener integral with respect to the two-sided Wiener process.

Proposition 4.2. *Let $(W_t, t \in \mathbb{R})$ be a Wiener process and let $H \in (0, 1)$. Define, for $t \geq 0$,*

$$X_t = \frac{1}{C(H)} \int_{\mathbb{R}} \left((t - s)_+^{H - \frac{1}{2}} - (-s)_+^{H - \frac{1}{2}} \right) dW_s \tag{4.5}$$

with

$$C(H)^2 = \int_{\mathbb{R}} \left((1-u)_+^{H-\frac{1}{2}} - (-u)_+^{H-\frac{1}{2}} \right)^2 du. \tag{4.6}$$

Then $(X_t, t \geq 0)$ is a standard fractional Brownian motion with Hurst index H.

Proof. First observe that every $t > 0$, the function

$$g_t(s) = (t-s)_+^{H-\frac{1}{2}} - (-s)_+^{H-\frac{1}{2}}$$

belongs to $L^2(\mathbb{R})$. This is due to the fact that $g_t(s)$ behaves as $(-s)^{H-\frac{3}{2}}$ as $-s$ goes to infinity. Then, for $0 \leq s \leq t$, we have, if X is given by (4.5),

$$\mathbf{E}|X_t - X_s|^2 = C(H)^{-2}\mathbf{E}\left(\int_{\mathbb{R}} \left((t-s)_+^{H-\frac{1}{2}} - (s-u)_+^{H-\frac{1}{2}} \right) dW_s \right)^2$$

$$= C(H)^{-2} \int_{\mathbb{R}} \left((t-u)_+^{H-\frac{1}{2}} - (s-u)_+^{H-\frac{1}{2}} \right)^2 du$$

and by the change of variables $t - u = -\tilde{u}$ and then $\frac{u}{t-s} = \tilde{u}$, we find

$$\mathbf{E}|X_t - X_s|^2 = C(H)^{-2} \int_{\mathbb{R}} \left((t-s-u)_+^{H-\frac{1}{2}} - (u)_+^{H-\frac{1}{2}} \right)^2 du$$

$$= C(H)^{-2}(t-s)^{2H} \int_{\mathbb{R}} \left((1-u)_+^{H-\frac{1}{2}} - (u)_+^{H-\frac{1}{2}} \right)^2 du$$

$$= (t-s)^{2H}$$

and this implies that X is a standard fractional Brownian motion because for every $s, t \geq 0$,

$$\mathbf{E}X_t X_s = \frac{1}{2} \left(\mathbf{E}X_t^2 + \mathbf{E}X_s^2 - \mathbf{E}(X_t - X_s)^2 \right)$$

$$= \frac{1}{2} \left(t^{2H} + s^{2H} - |t-s|^{2H} \right).$$

$\square$

Remark 4.2. One can show that the constant (4.6) is finite and it is equal to

$$\left(\int_0^\infty \left((1+s)^{H-\frac{1}{2}} - s^{H-\frac{1}{2}} \right)^2 ds + \frac{1}{2H} \right)^{\frac{1}{2}}.$$

It is also possible to represent the fractional Brownian motion as a Wiener integral with respect to the Wiener process on a finite interval.

Proposition 4.3. *Let $(B_t, t \in [0, T_0])$ be a Wiener process. If $H > \frac{1}{2}$, consider the kernel*

$$K_H(t, s) = c_H s^{\frac{1}{2} - H} \int_s^t (u - s)^{H - \frac{3}{2}} u^{H - \frac{1}{2}} du \qquad (4.7)$$

where $t > s$ and $c_H = \left(\frac{H(H-1)}{\beta(2 - 2H, H - \frac{1}{2})} \right)^{\frac{1}{2}}$. For $H < \frac{1}{2}$, define K_H by the following formula

$$K_H(t, s) = c_H \left[\left(\frac{t}{s} \right)^{H - \frac{1}{2}} (t - s)^{H - \frac{1}{2}} \right.$$

$$\left. - (H - \frac{1}{2}) s^{\frac{1}{2} - H} \int_s^t u^{H - \frac{3}{2}} (u - s)^{H - \frac{1}{2}} du \right], t > s \qquad (4.8)$$

with

$$c_H = \left(\frac{2H}{(1 - 2H)\beta(1 - 2H, H - \frac{1}{2})} \right)^{\frac{1}{2}}.$$

Then for every $t \in [0, T_0], H \neq \frac{1}{2}$, $K_H(t, \cdot)$ belongs to $L^2([0, T_0])$. For $t \in [0, T_0]$, let

$$B_t^H = \int_0^t K_H(t, s) dW_s.$$

Then $(B_t^H, t \in [0, T_0])$ is a fractional Brownian motion with Hurst parameter H.

Proof. It follows from the relation

$$\int_0^{t \wedge s} K_H(t, u) K_H(s, u) du = R_H(t, s), \qquad (4.9)$$

with R_H from (2.6). We refer to [Nualart (2006)] for a detailed proof. $\square$

Another interesting example of a process which can be expressed as a Wiener integral with respect to the Brownian motion is the Ornstein-Uhlenbeck process.

Example 4.2. Let $(B_t, t \geq 0)$ be a standard Brownian motion and $k > 0$. For every $t \geq 0$, set

$$X_t = \int_0^t e^{-k(t-s)} dW_s.$$

The process $(X_t, t \geq 0)$ is well-defined since

$$\mathbf{E} X_t^2 = \int_0^t e^{-2k(t-s)} ds = \frac{1}{2k} \left(1 - e^{-2kt} \right), \quad t \geq 0.$$

The process $(X_t, t \geq 0)$ is a centered Gaussian process with covariance given by, for $0 \leq s \leq t$,

$$\mathbf{E}X_t X_s = \int_0^{t \wedge s} e^{-k(t-u)} e^{-k(s-u)} du = \frac{1}{2k} e^{-k(t+s)} \left(e^{2ks} - 1 \right)$$

$$= \frac{1}{2k} \left(e^{-k(t-s)} - e^{-k(t+s)} \right),$$

so for $s, t \geq 0$,

$$\mathbf{E}X_t X_s = \frac{1}{2k} \left(e^{-k|t-s|} - e^{-k(t+s)} \right).$$

Example 4.3. Let $(W_t, t \in \mathbb{R})$ be two-sided Wiener process (see (3.4)) and $k > 0$. Define the process $(Y_t, t \geq 0)$ by

$$Y_t = \int_{-\infty}^t e^{-k(t-s)} dW_s, \quad t \geq 0.$$

Then for every $t \geq 0$,

$$\mathbf{E}Y_t^2 = -\int_{-\infty}^t e^{-2k(t-s)} ds = \frac{1}{2k}$$

and for $s, t \geq 0$,

$$\mathbf{E}Y_t Y_s = e^{-k(t+s)} \int_{-\infty}^{t \wedge s} e^{2ku} du = \frac{1}{2k} e^{-k|t-s|}.$$

It follows from Proposition 1.2 that the process Y is stationary. In particular, for every $t \geq 0$,

$$Y_t \sim N \left(0, \frac{1}{2k} \right).$$

The process $(Y_t, t \geq 0)$ is called the stationary Ornstein-Uhlenbeck process.

4.3 The fractional Brownian motion as an isonormal process

We define the canonical Hilbert space associated to the fBm in the following way. Let $(B_t^H, t \geq 0)$ be a standard fractional Brownian motion with Hurst parameter $H \in (0, 1)$. Consider $\mathcal{E}$ to be the linear space generated by the indicator functions $\{1_{[0,t]}, t > 0\}$ and define $\mathcal{H}$ to be the closure of $\mathcal{E}$ with respect to the inner product

$$\langle 1_{[0,t]}, 1_{[0,s]} \rangle_H = R_H(t, s)$$

with R_H from (2.6). Consider the mapping from $\mathcal{E}$ onto $L^2(\Omega)$ defined as

$$1_{[0,t]} \to B_t^H \text{ for any } t > 0. \qquad (4.10)$$

Since $\mathcal{E}$ is dense in $\mathcal{H}$, the mapping (4.10) can be extended to an isometry between $\mathcal{H}$ and $L^2(\Omega)$. Let $B^H(h)$ the image of $h \in \mathcal{H}$ through this isometry. Then we denote

$$B^H(h) = \int_0^\infty h(s) dB_s^H$$

and we call $B^H(h)$ the Wiener integral of $h \in \mathcal{H}$ with respect to the standard fractional Brownian motion. The process $(B^H(h), h \in \mathcal{H})$ is an isonormal process in the sense of Definition 4.1, since, for every $h, g \in \mathcal{E}$

$$\mathbf{E} B^H(h) B^H(g) = \langle h, g \rangle_{\mathcal{H}}$$

and by the density of $\mathcal{E}$ in $\mathcal{H}$, the above relation can be extended to every $h, g \in \mathcal{H}$.

We can define an associated canonical Hilbert space for $(B_t^H, t \in T)$ where T is any interval included in $\mathbb{R}$. The construction is similar. Consider first $\mathcal{E}$ to be the linear space generated by the indicator functions $\{1_{[a,b]}, a < b, a, b \in T\}$ and let $\mathcal{H}_T$ be the closure of $\mathcal{E}$ with respect to the inner product

$$\langle 1_{[a,b]}, 1_{[c,d]} \rangle_{\mathcal{H}_T} = R_H(b,d) - R_H(a,d) - R_H(c,b) + R_H(a,d)$$
$$= \frac{1}{2} \left(|a-d|^{2H} + |b-c|^{2H} - |b-d|^{2H} - |a-c|^{2H} \right).$$

Consider the mapping

$$1_{[a,b]} \to B_b^H - B_a^H : \mathcal{E} \to L^2(\Omega).$$

Via the density of $\mathcal{E}$ into $\mathcal{H}_T$, this mapping can be extended to an isometry between $\mathcal{H}_T$ and $L^2(\Omega)$. For $h \in \mathcal{H}_T$, $B^H(h)$ will be its image by this isometry and we denote again

$$B^H(h) = \int_T h(s) dB_s^H, \quad h \in \mathcal{H}_T.$$

The above integral is called the Wiener integral of $h \in \mathcal{H}_T$ with respect to $(B_t^H, t \in T)$ and the family $(B^H(h), h \in \mathcal{H}_T)$ becomes an isonormal process.

The canonical Hilbert space associated to B^H has the following properties (see e.g. [Pipiras and Taqqu (2001)] or [Nualart (2006)])

(1) If $H > \frac{1}{2}$, the Hilbert space $\mathcal{H}_T$ may contain distributions. It is therefore more practical to work with subspaces of $\mathcal{H}_T$ that are sets of functions. Such a subspace is

$$|\mathcal{H}_T| = \left\{ f : T \to \mathbb{R} \; \middle| \; \int_T \int_T |f(u)||f(v)||u - v|^{2H-2} dv du < \infty \right\}.$$

Then $|\mathcal{H}_T|$ is a strict subspace of $\mathcal{H}_T$ and if T is an interval with finite Lebesgue measure, we have the inclusions

$$L^2(T) \subset L^{\frac{1}{H}}(T) \subset |\mathcal{H}_T| \subset \mathcal{H}_T. \tag{4.11}$$

(2) The space $|\mathcal{H}_T|$ is not complete with respect to the norm $\|\cdot\|_{\mathcal{H}_T}$ but it is a Banach space with respect to the norm

$$\|f\|_{|\mathcal{H}_T|}^2 = \int_T \int_T |f(u)||f(v)||u - v|^{2H-2} dv du.$$

(3) If $H > \frac{1}{2}$ and f, g are two elements in the space $|\mathcal{H}_T|$, their scalar product in $\mathcal{H}_T$ can be expressed by

$$\langle f, g \rangle_{\mathcal{H}_T} = \alpha_H \int_T \int_T du dv |u - v|^{2H-2} f(u)g(v), \tag{4.12}$$

where $\alpha_H = H(2H - 1)$. In particular, we have the useful identity for $s, t \in T = [0, T_0]$,

$$R_H(t, s) = \alpha_H \int_0^t \int_0^s |u - v|^{2H-2} dv du. \tag{4.13}$$

(4) If $H < \frac{1}{2}$, then $\mathcal{H}_T$ is a space of functions. If $T \subset \mathbb{R}$ has finite Lebesque measure, we can prove the following inclusion

$$C^\gamma(T) \subset \mathcal{H}_T \subset L^2(T)$$

for every $\gamma > \frac{1}{2} - H$ where $C^\gamma(T)$ is the space of Hölder continuous functions on T of order γ.

Thus, if $f, g \in \mathcal{H}_T$, the Wiener integral with respect to $(B_t^H, t \in T)$ satisfies the Wiener isometry when $H > \frac{1}{2}$

$$\mathbf{E} \int_T f(s) dB_s^H \int_T g(s) dB_s^H = \langle f, g \rangle_{\mathcal{H}_T} \tag{4.14}$$

$$= \alpha_H \int_T \int_T du dv |u - v|^{2H-2} f(u)g(v).$$

As a first example, let us define the fractional Ornstein-Uhlenbeck process, which constitutes a fractional counterpart of the process defined in Example 4.2.

Example 4.4. Let $k > 0$ and let $(B_t^H, t \geq 0)$ be a standard fBm with $H \in \left(\frac{1}{2}, 1\right)$. Define

$$X_t = \int_0^t e^{-k(t-s)} dB_s^H, \quad t \geq 0. \tag{4.15}$$

The above Wiener integral with respect to the standard fBm B^H exists since for every $t \geq 0$,

$$\mathbf{E}X_t^2 = \alpha_H \int_0^t \int_0^t e^{-k(t-u)} e^{-k(t-v)} |u - v|^{2H} du\, dv$$

$$\leq \alpha_H \int_0^t \int_0^t |u - v|^{2H} du\, dv = t^{2H} < \infty$$

where the last equality is obtained via (4.12). The centered Gaussian process $(X_t, t \geq 0)$ given by (4.15) is called the fractional Ornstein-Uhlenbeck process. Its covariance is, for $s, t \geq 0$,

$$\mathbf{E}X_t X_s = \alpha_H e^{-k(t+s)} \int_0^t du \int_0^s dv\, e^{-k(u+v)} |u - v|^{2H}.$$

As for $H = \frac{1}{2}$, we can define a stationary fractional Ornstein-Uhlenbeck process.

Example 4.5. Let $(B_t^H, t \in \mathbb{R})$ be a fBm with $H \in \left(\frac{1}{2}, 1\right)$. Let $k > 0$. Define

$$Y_t = \int_{-\infty}^t e^{-k(t-s)} dB_s^H, \quad t \geq 0. \tag{4.16}$$

We notice that the process given by (4.16) is well-defined. For $t > 0$,

$$\mathbf{E}Y_t^2 = \alpha_H \int_{-\infty}^t \int_{-\infty}^t e^{-k(t-u)} e^{-k(t-v)} |u - v|^{2H-2} du\, dv$$

$$= \alpha_H \int_0^\infty \int_0^\infty e^{-ku} e^{-kv} |u - v|^{2H-2} du\, dv$$

$$= 2\alpha_H \int_0^\infty du\, e^{-ku} \int_0^u dv\, e^{-kv} (u - v)^{2H-2}$$

and by the change of variable $u - v = w$ in the integral dv,

$$\mathbf{E}Y_t^2 = 2\alpha_H \int_0^\infty du\, e^{-2ku} \int_0^u dw\, e^{kw} w^{2H-2}$$

$$= 2\alpha_H \int_0^\infty dw\, e^{kw} w^{2H-2} \int_w^\infty du\, e^{-2ku}$$

$$= \frac{2\alpha_H}{2k} \int_0^\infty dw\, e^{-kw} w^{2H-2} := C_0 < \infty.$$

Thus, for every $t \geq 0$, $Y_t \sim N(0, C_0)$. We can see that $(Y_t, t \geq 0)$ is a stationary Gaussian process because for every $h > 0$,

$$\mathbf{E}Y_{t+h}Y_{s+h} = \alpha_H \int_{-\infty}^{t+h} du \int_{-\infty}^{s+h} dv e^{-k(t+h-u)} e^{-k(s+h-v)} |u-v|^{2H-2}$$

$$= \alpha_H \int_{-\infty}^{t} \int_{-\infty}^{t} e^{-k(t-u)} e^{-k(t-v)} |u-v|^{2H-2} du dv$$

$$= \mathbf{E}Y_t Y_s$$

for every $s, t \geq 0$. So, by (1.1), the covariance of Y can be written as

$$\mathbf{E}Y_t Y_s = \mathbf{E}Y_0 Y_{t-s}$$

$$= \alpha_H \int_{-\infty}^{0} \int_{-\infty}^{t-s} e^{ku} e^{-k(t-s-v)} |u-|^{2H-2} dv du$$

$$= Q(t-s)$$

with

$$Q(x) = \alpha_H e^{-kx} \int_{-\infty}^{0} du \int_{-\infty}^{x} dv e^{k(u+v)} |u-v|^{2H-2}.$$

The process $(Y_t, t \geq 0)$ given by (4.16) is called the stationary fractional Ornstein-Uhlenbeck process.

The following result provides an interesting reproduction property of the fractional Brownian motion. It has first proved in [Jost (2006)].

Proposition 4.4. *Let* $K \in \left(\frac{1}{2}, 1\right)$ *and* $H \in (0, 1)$. *Let* $(B_t^K, t \geq 0)$ *be a fractional Brownian motion with Hurst index* K. *Set, for* $t \geq 0$,

$$B_t^H = \frac{1}{C(H, K)} \int_{\mathbb{R}} \left((t-u)_+^{H-K} - (-u)_+^{H-K} \right) dB_s^K \qquad (4.17)$$

with

$$C(H, K)^2 = \alpha_H \int_{\mathbb{R}} \int_{\mathbb{R}} du dv \left((1+u)_+^{H-K} - (u)_+^{H-k} \right)$$

$$\left((1+v)_+^{H-K} - (v)_+^{H-K} \right) |u-v|^{2K-2}.$$

Then $(B_t^H, t \geq 0)$ *is a fBm with Hurst parameter* H.

Proof. The Wiener integral in (4.17) exists because the function $u \to \left((t-u)_+^{H-K} - (-u)_+^{H-K} \right)$ (denoted $h_t(u)$) behaves as $(-u)^{H-K-1}$ when $-u$ goes to infinity and this implies that the above function satisfies

$$\int_{\mathbb{R}} \int_{\mathbb{R}} du dv |h_t(u)| |h_t(v)| |u-v|^{2K-2} < \infty.$$

Now, for $0 \le s \le t$, we have

$$\mathbf{E}|B_t^H - B_s^H|^2$$

$$= \frac{\alpha_H}{C(H,K)^2} \int_{\mathbb{R}} \int_{\mathbb{R}} du dv |u-v|^{2K-2}$$

$$\left((t-u)_+^{H-K} - (s-u)_+^{H-K}\right)\left((t-v)_+^{H-K} - (s-v)_+^{H-K}\right)$$

$$= \frac{\alpha_H}{C(H,K)^2} \int_{\mathbb{R}} \int_{\mathbb{R}} du dv |u-v|^{2K-2}$$

$$\left((t-s+u)_+^{H-K} - (u)_+^{H-K}\right)\left((t-s+v)_+^{H-K} - (v)_+^{H-K}\right)$$

$$= \frac{\alpha_H}{C(H,K)^2} |t-s|^{2(H-K)} \int_{\mathbb{R}} \int_{\mathbb{R}} du dv |u-v|^{2K-2}$$

$$\left(\left(1+\frac{u}{t-s}\right)_+^{H-K} - \left(\frac{u}{t-s}\right)_+^{H-K}\right)\left(\left(1+\frac{v}{t-s}\right)_+^{H-K} - \left(\frac{v}{t-s}\right)_+^{H-K}\right)$$

and with the change of variables $\frac{u}{t-s} = \tilde{u}, \frac{v}{t-s} = \tilde{v}$ we obtain

$$\mathbf{E}|B_t^H - B_s^H|^2$$

$$= \frac{\alpha_H}{C(H,K)^2} |t-s|^{2H} \int_{\mathbb{R}} \int_{\mathbb{R}} du dv$$

$$\left((1+u)_+^{H-K} - (u)_+^{H-K}\right)\left((1+v)_+^{H-K} - (v)_+^{H-K}\right)|u-v|^{2K-2}$$

$$= |t-s|^{2H}$$

which implies that B^H given by (4.17) is a H-fBm (i.e. a fBm with Hurst index H). □

4.4 The space-time white noise as an isonormal process

Let $(W(t,A), t \ge 0, A \in \mathcal{B}_b(\mathbb{R}^d))$ be a space-time white noise, i.e. a centered Gaussian process with covariance (3.5), see Definition 3.4. We can also associate to it an isonormal process via a construction similar to those above. We take $\mathcal{E}$ to be the vector space generated by

$$\{1_{[0,t] \times A}, t \ge 0, A \in \mathcal{B}_b(\mathbb{R}^d)\}$$

and we notice that this set is dense in $L^2([0,\infty) \times \mathbb{R}^d)$. We consider the mapping

$$1_{[0,t] \times A} \to W(t,A) : \mathcal{E} \to L^2(\Omega)$$

which is an isometry from $\mathcal{E}$ onto $L^2(\Omega)$ and we extend it to an isometry from $L^2([0,\infty) \times \mathbb{R}^d)$ to $L^2(\Omega)$. We set $W(\varphi)$ the image of $\varphi \in L^2([0,\infty) \times$

$\mathbb{R}^d$) via this application which will also be called the Wiener integral of φ with respect to the space-time white noise W. We also use the notation

$$W(\varphi) = \int_0^\infty \int_{\mathbb{R}^d} \varphi(s,y) W(ds,dy).$$

It then satisfies, for $\varphi_1, \varphi_2 \in L^2([0,\infty) \times \mathbb{R}^d)$,

$$\mathbf{E}\left(\int_0^\infty \int_{\mathbb{R}^d} \varphi_1(s,y) W(ds,dy)\right) \left(\int_0^\infty \int_{\mathbb{R}^d} \varphi_2(s,y) W(ds,dy)\right)$$
$$= \int_0^\infty \int_{\mathbb{R}^d} \varphi_1(s,y) \varphi_2(s,y) dyds = \langle \varphi_1, \varphi_2 \rangle_{L^2([0,\infty)\times\mathbb{R}^d)}. \qquad (4.18)$$

Therefore $(W(\varphi), \varphi \in L^2([0,\infty) \times \mathbb{R}^d))$ is an isonormal process, as in Definition 4.1.

We give some examples of Wiener integrals with respect to the space-time white noise.

Example 4.6. Consider the function f defined on $\mathbb{R}_+ \times \mathbb{R}$ given by

$$f(s,y) = 1_{[0,T_1]}(s) e^{-(T_1-s)} 1_{[0,T_2]}(y) e^{-(T_2-y)} \qquad (4.19)$$

with $T_1, T_2 > 0$ fixed. Then $f \in L^2(\mathbb{R}_+ \times \mathbb{R})$ because

$$\int_0^\infty \int_{\mathbb{R}} f^2(s,y) dyds = \int_0^{T_1} \int_0^{T_2} e^{-2(T_1-s)} e^{-2(T_2-y)} dyds$$
$$= \frac{1}{4}\left(1 - e^{-2T_1}\right)\left(1 - e^{-2T_2}\right) < \infty.$$

The process $(X_{T_1,T_2}, T_1, T_2 \geq 0)$ given by

$$X_{T_1,T_2} = \int_0^{T_1} \int_0^{T_2} e^{-(T_1-s)} e^{-(T_2-y)} W(ds,dy)$$

is a two-parameter Ornstein-Uhlenbeck process (there are actually several ways to define the Ornstein-Uhlenbeck sheet).

Example 4.7. Let, for $t > 0, x \in \mathbb{R}^d$,

$$G(t,x) = \frac{1}{(2\pi t)^{\frac{d}{2}}} e^{-\frac{|x|^2}{2t}}$$

the Gaussian kernel of variance $t > 0$. Then the mapping

$$(s,y) \to G(t-s, x-y)$$

belongs to $L^2((0,\infty) \times \mathbb{R}^d)$ if and only if $d = 1$ (see Section 5.1.1 for the proof). Consequently the Wiener integral

$$X(t,x) = \int_0^\infty \int_{\mathbb{R}^d} G(t-s, x-y) W(ds,dy)$$

is well-defined if and only if $d = 1$. The process $(X(t,x), t \geq 0, x \in \mathbb{R})$ constitutes the mild solution to the stochastic heat equation studied in details in the next chapters.

Obviously, due to the correspondence between the anisotropic fractional Brownian sheet and the space-time white noise discussed in Section 3.2, the Wiener integral can be defined with respect to the anisotropic Brownian sheet $(W(x), x \in T \subset \mathbb{R}^d)$. This random field generates an isonormal process $(W(h), h \in L^2(T))$ and $W(h)$ will be the Wiener integral of $h \in L^2(T)$ with respect to the Brownian sheet.

Let us give the Wiener integral representation of the fractional Brownian sheet.

Proposition 4.5. *Let $(W(x), x \in \mathbb{R}^d)$ be an anisotropic Brownian sheet and define, for $t = (t_1, \ldots, t_d) \in \mathbb{R}^d_+$ and for $H = (H_1, \ldots, H_d) \in (0,1)^d$,*

$$W_t^H = \frac{1}{C(H_1) \ldots C(H_d)} \int_{\mathbb{R}^d} \left(\prod_{i=1}^d \left((t_i - u_i)_+^{H_i - \frac{1}{2}} - (-u_i)_+^{H_i - \frac{1}{2}} \right) \right) W(du)$$

where $W(du)$ means $W(du_1, \ldots, du_d)$ and $C(H_i)$ is the constant given by (4.6). Then the random field $(W_t^H, t \in \mathbb{R}^d_+)$ is an anisotropic fractional Brownian sheet with Hurst parameter $H = (H_1, \ldots, H_d) \in (0,1)^d$.

Proof. To see it, we notice that W^H is a centered d-parameter Gaussian process and it suffices to calculate its covariance. For $s = (s_1, \ldots, s_d), t = (t_1, \ldots, t_d) \in (0, \infty)^d$, we have

$$\mathbf{E} W_t^H W_s^H = \frac{1}{C(H_1)^2 \ldots C(H_d)^2} \int_{\mathbb{R}^d} du_1 \ldots du_d$$
$$\times \left[\prod_{i=1}^d \left((t_i - u_i)_+^{H_i - \frac{1}{2}} - (-u_i)_+^{H_i - \frac{1}{2}} \right) \right]^2$$
$$= \prod_{i=1}^d \left[\frac{1}{C(H_i)^2} \int_{\mathbb{R}} \left((t_i - u_i)_+^{H_i - \frac{1}{2}} - (-u_i)_+^{H_i - \frac{1}{2}} \right)^2 du_i \right]$$
$$= \prod_{i=1}^d R_{H_i}(t_i, s_i)$$

where for $i = 1, \ldots, d$, R_{H_i} is the covariance of the fBm with Hurst index $H_i \in (0,1)$ and the last equality above follows from Proposition 4.2. $\square$

It is also possible to give a Wiener integral representation of the anisotropic fractional Brownian sheet on the half-real line.

Proposition 4.6. *Let $(B_t, t \in \mathbb{R}^d_+)$ be an anisotropic d-parameter standard Wiener process and define*

$$W_t^H = \int_{[0,t]} K_{H_1}(t_1, u_1) \ldots K_{H_d}(t_d, u_d) W(du_1, \ldots, du_d)$$

for $H = (H_1, \ldots, H_d) \in (0,1)^d$ and $t = (t_1, \ldots, t_d) \in (0, \infty)^d$ with the kernel K_H given by (4.7) or (4.8). Then $(W_t^H, t \in [0, \infty)^d)$ is an anisotropic fractional Brownian sheet.

Proof. By formula (4.9) in Proposition 4.3, for $s, t \in [0, \infty)^d$,

$$\mathbf{E} W_t^H W_s^H$$

$$= \int_{[0,t] \cap [0,s]} K_{H_1}(t_1, u_1) \ldots K_{H_d}(t_d, u_d) K_{H_1}(s_1, u_1) \ldots K_{H_d}(s_d, u_d) du_1 \ldots du_d$$

$$= \prod_{i=1}^{d} \int_{[0,t_i]} K_{H_1}(t_i, u_i) K_{H_1}(s_i, u_i) du_i = \prod_{i=1}^{d} R_{H_i}(t_i, s_i).$$

$\square$

Concerning the integral representation of the isotropic fractional Brownian sheet, we have the following result. Recall that the isotropic Wiener process is defined by (3.9).

Proposition 4.7. *Let $(W(x), x \in \mathbb{R}^d)$ be an anisotropic Brownian sheet and define, for $t \in [0, \infty)^d$ and $H \in (0, 1)$*

$$X_t^H = C(H, d) \int_{\mathbb{R}^d} \left[\|t - u\|^{H - \frac{d}{2}} - \| - u\|^{H - \frac{d}{2}} \right] W(du)$$

with

$$C(H, d)^2 = \left(\int_{\mathbb{R}^d} \left[\|e_1 - w\|^{H - \frac{d}{2}} - \|w\|^{H - \frac{d}{2}} \right]^2 dw \right)^{-1}$$

where $e_1 = (1, 0, \ldots, 0) \in \mathbb{R}^d$ and $|\cdot|$ is the Euclidean norm. Then the process $(X_t^H, t \in \mathbb{R}_+^d)$ is an isotropic fractional Brownian sheet.

Proof. Let $s, t \in \mathbb{R}_+^d$. Then

$$X_t^H - X_s^H = C(H, d) \int_{\mathbb{R}^d} \left[\|t - u\|^{H - \frac{d}{2}} - \|s - u\|^{H - \frac{d}{2}} \right] W(du)$$

and

$$\mathbf{E} |X_t^H - X_s^H|^2 = C(H, d)^2 \int_{\mathbb{R}^d} \left[\|t - u\|^{H - \frac{d}{2}} - \|s - u\|^{H - \frac{d}{2}} \right]^2 du$$

$$= C(H, d)^2 \int_{\mathbb{R}^d} \left[\|t - s - u\|^{H - \frac{d}{2}} - \|u\|^{H - \frac{d}{2}} \right]^2 du.$$

We perform the change of variables $v = \varphi(u)$ who maps the canonical basis $(e_1, \ldots, e_d)$ of $\mathbb{R}^d$ into a basis $\left(\frac{t-s}{\|t-s\|}, f_2, \ldots, f_d \right)$. Since φ is orthogonal,

$$J\varphi^{-1}(v) = |\det(\varphi^{-1})| = 1.$$

We can also write, for every $t, s, u \in \mathbb{R}_+^d$

$$
\begin{aligned}
\|t - s - u\|^2 &= \|t - s\|^2 + \|u\|^2 - 2\langle t - s, u \rangle \\
&= \|t - s\|^2 + \|u\|^2 - 2\|t - s\|v_1 \\
&= \|\ \|t - s\|e_1 - v\|^2
\end{aligned}
$$

if $v = (v_1, \ldots, v_d)$. Above we used

$$
\langle t - s, u \rangle = \langle \varphi^{-1}(t - s), v \rangle = \langle \|t - s\|e_1, v \rangle = \|t - s\|v_1.
$$

So

$$
\begin{aligned}
&\mathbf{E}|X_t^H - X_s^H|^2 \\
&= C(H, d)^2 \int_{\mathbb{R}^d} \left[\|\ \|t - s\|e_1 - v\|^{H - \frac{d}{2}} - \|v\|^{H - \frac{d}{2}} \right]^2 du \\
&= C(H, d)^2 \|t - s\|^{2H - d} \int_{\mathbb{R}^d} \left[\left\| e_1 - \frac{v}{\|t - s\|} \right\|^{H - \frac{d}{2}} - \left| \frac{v}{\|t - s\|} \right|^{H - \frac{d}{2}} \right]^2 du.
\end{aligned}
$$

Finally, by the change of variables $w = \frac{v}{\|t - s\|}$, we will have, for $s, t \in [0, \infty)^d$,

$$
\begin{aligned}
\mathbf{E}|X_t^H - X_s^H|^2 &= C(H, d)^2 \|t - s\|^{2H} \int_{\mathbb{R}^d} \left[\|e_1 - w\|^{H - \frac{d}{2}} - \|w\|^{H - \frac{d}{2}} \right]^2 dw \\
&= \|t - s\|^{2H}.
\end{aligned}
$$

This implies that the covariance of X^H is given by (3.8). $\qquad\square$

4.5 The white-colored noise as an isonormal process

Recall that the white-colored noise is a centered Gaussian field indexed by $T = \mathbb{R}_+ \times \mathcal{B}_b(\mathbb{R}^d)$ with covariance given by (3.12). This random field can be also viewed as an isonormal process. As before we take $\mathcal{E}$ to be the vector space generated by $\left(1_{[0,t] \times A}, t > 0, A \in \mathcal{B}_b(\mathbb{R}^d) \right)$. We define the Hilbert space $\mathcal{P}$ as the completion of $\mathcal{E}$ with respect to the inner product

$$
\langle 1_{[0,t] \times A}, 1_{[0,s] \times B} \rangle_{\mathcal{P}} = (t \wedge s) \int_A \int_B f(x - y) dx dy \qquad (4.20)
$$

for $t, s \geq 0$, $A, B \in \mathcal{B}_b(\mathbb{R}^d)$, with f the Riesz kernel (3.13) or order $\gamma \in (0, d)$. The mapping

$$
1_{[0,t] \times A} \to W^\gamma(t, A) : \mathcal{E} \to L^2(\Omega)
$$

is then extended to an isometry between $\mathcal{P}$ and $L^2(\Omega)$ and for $\varphi \in \mathcal{P}$, $W^\gamma(\varphi)$ will be its image via this mapping. In this way $(W^\gamma(\varphi), \varphi \in \mathcal{P})$ becomes an isonormal process since, for every $\varphi_1, \varphi_2 \in \mathcal{E}$,

$$\mathbf{E}W^\gamma(\varphi_1)W^\gamma(\varphi_2) = \langle \varphi_1, \varphi_2 \rangle_\mathcal{P}$$

and this relation can be extended, by density, to every $\varphi_1, \varphi_2 \in \mathcal{P}$. We also write

$$W^\gamma(\varphi) = \int_0^\infty \int_{\mathbb{R}^d} \varphi(s,y) W^\gamma(ds, dy)$$

to indicate the Wiener integral of $\varphi \in \mathcal{P}$ with respect to the white-colored noise W^γ.

We know the following about the space $\mathcal{P}$ (see [Dalang (1999)] or [Balan and Tudor (2008)]).

(1) The space $\mathcal{P}$ is not a space of functions and it may contain distributions.
(2) A subset of $\mathcal{P}$ which is a space of functions is the space $|\mathcal{P}|$ which is defined as the set of $\varphi : \mathbb{R}_+ \times \mathbb{R}^d \to \mathbb{R}$ such that

$$\int_0^\infty \int_{\mathbb{R}^d} \int_{\mathbb{R}^d} |\varphi(s,y)| \cdot |\varphi(s,z)| f(y-z) \, dz \, dy \, ds < \infty. \qquad (4.21)$$

(3) If $\varphi, \psi \in |\mathcal{P}|$, then their scalar product in $\mathcal{P}$ can be expressed as

$$\langle \varphi, \psi \rangle_\mathcal{P} = \int_0^\infty \int_{\mathbb{R}^d} \varphi(s,y) \varphi(s,z) f(y-z) \, dz \, dy \, ds$$

and consequently, the isometry of the Wiener integral with respect to W^γ reads

$$\mathbf{E}W^\gamma(\varphi)W^\gamma(\psi) = \int_0^\infty \int_{\mathbb{R}^d} \varphi(s,y) \varphi(s,z) f(y-z) \, dz \, dy \, ds \qquad (4.22)$$

if $\varphi, \psi \in |\mathcal{P}|$.

Let us consider an example of a Wiener integral with respect to the white-colored noise.

Example 4.8. Let W^γ be a white-colored noise with $d = 1$ and $\gamma \in (0,1)$ and let g be the function given by (4.19). Then $g \in |\mathcal{P}|$ since, with $g_{\gamma,1}$

from (3.13),

$$\int_0^\infty \int_{\mathbb{R}} \int_{\mathbb{R}} |g(s,y)||g(s,z)|f(y-z)$$

$$= g_{\gamma,1} \int_0^\infty \int_{\mathbb{R}} \int_{\mathbb{R}} |f(s,y)||f(s,z)||y-z|^{\gamma-1} dy dz ds$$

$$= g_{\gamma,1} \int_0^{T_1} ds \int_0^{T_2} dy \int_0^{T_2} dz \, e^{-(T_1-s)} e^{-(T_2-y)} e^{-(T_2-z)} |y-z|^{\gamma-1}$$

$$\leq g_{\gamma,1} T_1 \int_0^{T_2} dy \int_0^{T_2} dz |y-z|^{\gamma-1},$$

and the above integral if finite since $\gamma > 0$. The process $(Y_{T_1,T_2}, T_1, T_2 \geq 0)$ defined as

$$Y_{T_1,T_2} = \int_0^{T_1} \int_0^{T_2} e^{-(T_1-s)} e^{-(T_2-y)} W^\gamma(ds, dy)$$

can be viewed as a white-colored Ornstein-Uhlenbeck process (OU), i.e. it behaves as the standard OU process in time and the fractional OU process in space.

PART 2
Stochastic heat and wave equations with additive Gaussian noise

Chapter 5

The stochastic heat equation with space-time white noise

The heat equation is one of the most famous stochastic partial differential equations (SPDEs in the sequel). The simplest (deterministic) heat equation on an opet subset U of $\mathbb{R}^d$ with $d \geq 1$ can be written as

$$\frac{\partial u}{\partial t} = \Delta u(t, x), \quad t \geq 0, x \in U \tag{5.1}$$

where Δ stands for the standard Laplacian operator on $\mathbb{R}^d$, i.e. if $F : U \subset \mathbb{R}^d \to \mathbb{R}$, then

$$\Delta F(x) = \sum_{i=1}^{d} \frac{\partial^2 F}{\partial x_i^2}(x), \quad x = (x_1, \ldots, x_d) \in U. \tag{5.2}$$

A solution to the heat equation is a function u that satisfies (5.1). We call $t \geq 0$ the temporal (or time) variable and $x \in U \subset \mathbb{R}$ is the spatial variable. This equation describes the flow of the heat in a "standard" (homogenous and isotropic) medium, $u(t, x)$ being the temperature at time t and at point $x \in U$. If the medium is not standard, then the heat equation is perturbed by other factors. The stochastic heat equation describes the heat flow in a random medium and the random perturbation is often described by a space-time white noise (the random field presented in Section 3.2) which behaves as the Brownian motion both in time and in space. Sometimes, depending on the particularity of the random medium, it can be modelled by a noise with spatial (or temporal) correlation. When the medium is non-homogenous, the standard Laplacian operator in (5.1) is replaced by the fractional Laplacian, leading to the fractional stochastic heat equation (discussed in Sections 5.2 and 6.2).

We will discuss each of these situations: the heat equation (5.1) driven by a space-time white noise or by a white-colored Gaussian noise, and the fractional heat equation, i.e. the standard Laplacian in (5.1) is replaced by the fractional Laplacian. For each of these cases, we analyze the pathwise properties of the solution and its probability distribution. We will focus on the link between the solution to (5.1) and the fractional or bifractional Brownian motion.

5.1   The standard stochastic heat equation with space-time white noise

5.1.1   *General properties of the solution*

We consider first the following SPDE

$$\frac{\partial}{\partial t} u(t, x) = \frac{1}{2} \Delta u(t, x) + \dot{W}(t, x), \quad t \geq 0, x \in \mathbb{R}^d. \tag{5.3}$$

with vanishing initial condition $u(0, x) = 0$ for every $x \in \mathbb{R}^d$. In (5.38), Δ denotes the Laplacian and W is space-time white noise, i.e. $W = \big(W(t, A), t \geq 0, A \in \mathcal{B}(\mathbb{R}^d)\big)$ is a centered Gaussian field with covariance

$$\mathbf{E}W(t, A)W(s, B) = (t \wedge s)\lambda(A \cap B) \tag{5.4}$$

where λ denotes the Lebesgue measure on $\mathbb{R}^d$. The properties of the random field W with covariance (5.4) have been discussed in Section 3.2. In particular, it can be regarded as an anisotropic Brownian sheet, see Section 3.2.1. It is called space-time white noise because it behaves as a Wiener process both in respect to its temporal and spatial variables.

Let G be the following kernel

$$G(t, x) = \begin{cases} (2\pi t)^{-d/2} \exp\left(-\frac{\|x\|^2}{2t}\right) & \text{if } t > 0, x \in \mathbb{R}^d \\ 0 & \text{if } t \leq 0, x \in \mathbb{R}^d. \end{cases} \tag{5.5}$$

The kernel G is called the Green kernel, or the fundamental solution, associated to the heat equation. In particular, it solves

$$\frac{\partial}{\partial t} u(t, x) = \frac{1}{2} \Delta u(t, x)$$

for $t \geq 0, x \in \mathbb{R}^d$. Another property of the Green kernel, which serves as motivation for the definition of the mild solution to the heat equation, is the one given in the below remark.

Remark 5.1. Let G be the Green kernel (5.5) and take $f \in C^{1,2}([0, \infty) \times \mathbb{R}^d)$. Define

$$u(t, x) = \int_0^t ds \int_{\mathbb{R}^d} dy\, G(t - s, x - y)f(s, y). \tag{5.6}$$

Then one can show that the function u given by (5.6) solves the partial differential equation

$$\frac{\partial}{\partial t} u(t,x) = \frac{1}{2}\Delta u(t,x) + f(t,x)$$

for $t \geq 0, x \in \mathbb{R}^d$.

This motivates the definition of the mild solution (5.7) to the stochastic heat equation by replacing $f(s,y)dsdy$ in (5.6) by $\dot{W}(s,y)dsdy$, which is "stochastically" interpreted as $W(ds,dy)$. Another motivation to define the solution to the stochastic heat equation by (5.7) is its interpretation as a weak solution, see Proposition 5.3 below.

Let us define the concept of solution to (5.3). Let W be a space-time white noise (i.e. a centered Gaussian field with covariance (5.4)) and let, for every $t \geq 0$ and $x \in \mathbb{R}^d$,

$$u(t,x) = \int_0^t \int_{\mathbb{R}^d} G(t-s, x-y)W(ds,dy) \tag{5.7}$$

This is usually called the *mild* or (*evolutive*) solution to (5.3). The Wiener integral in (5.7) is a Wiener integral with respect to the space-time white noise, as defined in Section 4.4. We say that the mild solution exists if the Wiener integral in (5.7) is well-defined, i.e. if for every $t > 0, x \in \mathbb{R}^d$, the function

$$(s,y) \to G(t-s, x-y)1_{[0,t]}(s) : \mathbb{R}_+ \times \mathbb{R}^d$$

is square-integrable. The starting point of the analysis of the mild solution is to give a necessary and sufficient condition for its existence. Actually, we will see that the Wiener integral in (5.7) exists if and only if $d = 1$.

Proposition 5.1. *Let* $(u(t,x), t \geq 0, x \in \mathbb{R}^d)$ *be given by (5.7). Then u is a well-defined centered Gaussian random field if and only if $d = 1$. In this case, for every $t > 0, x \in \mathbb{R}$,*

$$u(t,x) \sim N\left(0, \frac{\sqrt{t}}{\sqrt{\pi}}\right)$$

and for every $p \geq 1, T > 0$,

$$\sup_{t \in [0,T], x \in \mathbb{R}} \mathbf{E}|u(t,x)|^p \leq C_T \tag{5.8}$$

with $C_T > 0$.

Proof. Indeed, for every $t \geq 0$ and $x \in \mathbb{R}^d$ we have

$$
\mathbf{E}u(t,x)^2 = \int_0^t ds \int_{\mathbb{R}^d} dy G(t-s, x-y)^2
$$

$$
= \int_0^t ds \int_{\mathbb{R}^d} dy \left(\frac{1}{(2\pi(t-s))^{\frac{d}{2}}} e^{-\frac{|x-y|^2}{2(t-s)}} \right)^2
$$

$$
= \int_0^t ds \int_{\mathbb{R}^d} dy \frac{1}{(2\pi s)^d} e^{-\frac{|y|^2}{s}} = \pi^{-\frac{d}{2}} 2^{-d} \int_0^t ds \ s^{-\frac{d}{2}}.
$$

The integral $\int_0^t s^{-\frac{d}{2}} ds$ is finite if and only if $d = 1$ and in this case

$$
\mathbf{E}u(t,x)^2 = \frac{\sqrt{t}}{\sqrt{\pi}}.
$$

Thus $u(t,x) \sim N\left(0, \frac{\sqrt{t}}{\sqrt{\pi}}\right)$ and for $t \in [0,T]$ and $x \in \mathbb{R}$,

$$
\mathbf{E}|u(t,x)|^p = \left(\frac{t}{\pi}\right)^{\frac{p}{4}} \mathbf{E}|Z|^p \leq \left(\frac{T}{\pi}\right)^{\frac{p}{4}} \mathbf{E}|Z|^p
$$

for every $p \geq 1$, where Z is a standard normal random variable. This gives
(5.8). $\qquad\qquad\square$

Hence, for the analysis of the stochastic heat equation driven by the space-time white noise, we have to restrict to the situation when the spatial dimension is $d = 1$. This will be overcame by considering a correlated noise in space, as discussed in the next chapters.

We will use the Fourier transform of the Green kernel (5.5). See [Treves (1975)] for a detailed presentation of the definition and properties of the Fourier transform. Let $\mathcal{S}(\mathbb{R}^d)$ be the Schwarz space of rapidly decreasing C^∞ functions on $\mathbb{R}^d$. For $f \in \mathcal{S}(\mathbb{R}^d)$ we define its Fourier transform by

$$
(\mathcal{F}f)(\xi) = \int_{\mathbb{R}^d} e^{i\langle x,\xi \rangle} dx, \quad \xi \in \mathbb{R}^d. \tag{5.9}
$$

The Fourier transform of f is also well-defined if f belongs to $L^1(\mathbb{R}^d)$ or $L^2(\mathbb{R}^d)$. We will need the Parseval-Plancherel identity

$$
\langle f, g \rangle_{L^2(\mathbb{R}^d)} = (2\pi)^{-d} \int_{\mathbb{R}^d} (\mathcal{F}f)(\xi) \overline{\mathcal{F}f}(\xi) d\xi, \tag{5.10}
$$

for any $f, g \in L^2(\mathbb{R}^d)$.

If G is the Green kernel given by (5.5), its Fourier transform with respect to the spatial variable (i.e. the Fourier transform of the function $x \to G(t,x)$) is given, for every $t > 0$, by

$$\mathcal{F}G(t,\cdot)(\xi) = (2\pi t)^{-d/2} \int_{\mathbb{R}^d} e^{i\langle x,\xi\rangle} \exp\left(-\frac{\|x\|^2}{2t}\right) dx = e^{-\frac{t\|\xi\|^2}{2}}, \qquad \xi \in \mathbb{R}^d. \tag{5.11}$$

Let us compute the covariance of the Gaussian random field u.

Proposition 5.2. *Let $(u(t,x), t \geq 0, x \in \mathbb{R})$ be given by (5.7). For every $s, t \geq 0$ with $s \leq t$ and for every $x, y \in \mathbb{R}$,*

$$\mathbf{E}u(t,x)u(s,y)$$

$$= \frac{1}{\sqrt{2\pi}} \left(\sqrt{t+s}\, e^{-\frac{|y-x|^2}{2(t+s)}} - \sqrt{t-s}\, e^{-\frac{|y-x|^2}{2(t-s)}} \right) - \frac{1}{\sqrt{\pi}}|y-x| \int_{\frac{|y-x|}{\sqrt{2(t+s)}}}^{\frac{|y-x|}{\sqrt{2(t-s)}}} e^{-z^2} dx$$

$$= \frac{1}{\sqrt{2\pi}} \left(\sqrt{t+s}\, e^{-\frac{|y-x|^2}{2(t+s)}} - \sqrt{t-s}\, e^{-\frac{|y-x|^2}{2(t-s)}} \right)$$

$$- \sqrt{2}|x-y|\,\mathrm{erf}\left(\frac{|x-y|}{\sqrt{2(t-s)}}\right) + \sqrt{2}|x-y|\,\mathrm{erf}\left(\frac{|x-y|}{\sqrt{2(t+s)}}\right) \tag{5.12}$$

where erf denotes the error function

$$\mathrm{erf}(x) = \frac{1}{\sqrt{2\pi}} \int_0^x e^{-z^2} dz. \tag{5.13}$$

Proof. We have, via Parseval's relation (5.10), for $0 \leq s \leq t, x, y \in \mathbb{R}$,

$$\mathbf{E}u(t,x)u(s,y) = \int_0^{t\wedge s} du \int_{\mathbb{R}} dz\, G(t-u, x-z)G(s-u, s-u)$$

$$= \frac{1}{2\pi} \int_0^{t\wedge s} du \int_{\mathbb{R}} d\xi\, e^{-i\xi(x-y)} \mathcal{F}G(t-u,\cdot)(\xi)\overline{\mathcal{F}G(s-u,\cdot)}(\xi)$$

and by (5.11),

$$\mathbf{E}u(t,x)u(s,y) = \frac{1}{2\pi} \int_0^{t\wedge s} du \int_{\mathbb{R}} d\xi\, e^{-i\xi(x-y)} e^{-\frac{1}{2}(t+s-2u)|\xi|^2}$$

$$= \frac{1}{\sqrt{2\pi}} \int_0^{t\wedge s} (t+s-2u)^{-\frac{1}{2}} e^{-\frac{|x-y|^2}{2(t+s-2u)}} du.$$

Assume $0 \leq s \leq t$. By using successively the change of variables $a = t+s-2u$ and $b = \frac{(y-x)^2}{2a}$, we have

$$\mathbf{E}u(t,x)u(s,y) = \frac{1}{2\sqrt{2\pi}} \int_{t-s}^{t+s} a^{-\frac{1}{2}} e^{-\frac{(y-x)^2}{2a}} da = \frac{1}{4\sqrt{\pi}}|y-x| \int_{\frac{(y-x)^2}{2(t+s)}}^{\frac{(y-x)^2}{2(t-s)}} b^{-\frac{3}{2}} e^{-b} db.$$

Next we integrate by parts and we obtain

$$\mathbf{E}u(t,x)u(s,y) = \frac{1}{\sqrt{2\pi}}\left[\sqrt{t+s}e^{-\frac{(y-x)^2}{2(t+s)}} - \sqrt{t-s}e^{-\frac{(y-x)^2}{2(t-s)}}\right] \quad (5.14)$$

$$-\frac{1}{2\sqrt{\pi}}|y-x|\int_{\frac{(y-x)^2}{2(t+s)}}^{\frac{(y-x)^2}{2(t-s)}} b^{-\frac{1}{2}}e^{-b}db.$$

To obtain the conclusion, we perform the change of variables $b = z^2$ in the last integral. □

The covariance (5.12) determines the law of the Gaussian random field (5.7). On the other hand, this covariance formula is pretty complex. We will see that, when the temporal or the spatial variable is fixed, the formula (5.12) reduces to a simpler expression.

5.1.2   *Relation with the weak solution*

Another possibility to define the solution to (5.3) is via *the weak formulation*. Let us discuss the relation between the mild solution (5.7) and the weak solution of the heat equation with space-time white noise. We recall the weak formulation of the solution to the heat equation, which is motivated by the integration by parts.

Definition 5.1. A stochastic process $(u(t,x), t \geq 0, x \in \mathbb{R})$ is a weak solution to (5.3) if

$$\int_0^T \int_{\mathbb{R}} u(t,x)\left(\frac{\partial \varphi}{\partial t} + \frac{\partial^2 \varphi}{\partial x^2}\right)(t,x)dxdt = -\int_0^T \int_{\mathbb{R}} \varphi(s,y)W(ds,dy)$$

for every $T > 0$ and for every test function $\varphi \in C^\infty([0,\infty) \times \mathbb{R})$ with compact support in $\mathbb{R}$ such that $\varphi(T,x) = 0$ for every $x \in \mathbb{R}$.

Proposition 5.3. *A process u is a mild solution to (5.3) if and only if it is a weak solution to (5.3).*

Proof. Suppose u is a mild solution given by (5.7) and consider a test function $\varphi \in C^\infty([0,\infty) \times \mathbb{R})$ with compact support in $\mathbb{R}$ such that $\varphi(T,x) =$

0 for every $x \in \mathbb{R}$. Then, by using Fubini's theorem,

$$
\int_0^T \int_{\mathbb{R}} u(t,x) \left(\frac{\partial \varphi}{\partial t} + \frac{\partial^2 \varphi}{\partial x^2} \right) (t,x) dx dt
$$

$$
= \int_0^T \int_{\mathbb{R}} \left(\int_0^t \int_{\mathbb{R}} G(t-s, x-y) W(ds,dy) \right) \left(\frac{\partial \varphi}{\partial t} + \frac{\partial^2 \varphi}{\partial x^2} \right) (t,x) dx dt
$$

$$
= \int_0^T \int_{\mathbb{R}} W(ds,dy) \int_s^T dt \int_{\mathbb{R}} dx
$$

$$
\times G(t-s, x-y) \left(\frac{\partial \varphi}{\partial t} + \frac{\partial^2 \varphi}{\partial x^2} \right) (t,x) \tag{5.15}
$$

and then we use the integration by parts to get

$$
\int_s^T dt \int_{\mathbb{R}} dx G(t-s, x-y) \frac{\partial \varphi}{\partial t}(t,x)
$$

$$
= -\varphi(s,y) - \int_s^T dt \int_{\mathbb{R}} dx \frac{\partial}{\partial t} G(t-s, x-y) \varphi(t,x)
$$

(see [Evans (2010)], pages 50-51 for a detailed proof, here we use the fact that $\frac{1}{\sqrt{2\pi(t-s)}} e^{-\frac{(-y)^2}{2(t-s)}}$ converges, as $t - s \to 0$, to the Dirac distribution $\delta_0(x-y)$) and

$$
\int_s^T dt \int_{\mathbb{R}} dx G(t-s, x-y) \frac{\partial^2 \varphi}{\partial x^2}(t,x) = \int_s^T dt \int_{\mathbb{R}} dx \frac{\partial^2}{\partial x^2} G(t-s, x-y) \varphi(t,x).
$$

Thus

$$
\int_s^T dt \int_{\mathbb{R}} dx G(t-s, x-y) \left(\frac{\partial \varphi}{\partial t} + \frac{\partial^2 \varphi}{\partial x^2} \right) (t,x) = -\varphi(s,y). \tag{5.16}
$$

By plugging (5.16) into (5.15) we deduce that u is a weak solution. For the proof of the converse direction we refer to [Folland (1975)] or [Evans (2010)]. □

5.1.3   *Analysis of the solution in time*

Now, the purpose is to study the behavior of the mild solution (5.7) with respect to the temporal variable, i.e. to study the (one-parameter) Gaussian process $(u(t,x), t \geq 0)$ with $x \in \mathbb{R}$ fixed. From the covariance formula (5.12) we obtain immediately the temporal covariance of the solution u. This expression of the temporal covariance will give the distribution of the

solution in time and many of its trajectorial and distributional properties
will follow.

Proposition 5.4. *Let $(u(t,x), t \geq 0, x \in \mathbb{R})$ be given by (5.7). For every $s, t \geq 0$ and for every $x \in \mathbb{R}$, we have*

$$\mathbf{E}u(t,x)u(s,x) = \frac{1}{\sqrt{2\pi}}\left(\sqrt{t+s} - \sqrt{|t-s|}\right). \tag{5.17}$$

Proof. Take $x = y$ in (5.12). $\qquad\square$

Remark 5.2. In particular, for every $t > 0, x \in \mathbb{R}$,

$$u(t,x) \sim N\left(0, \sqrt{\frac{t}{\pi}}\right),$$

as in Proposition 5.1.

We obtain the following connection between the solution to the heat
equation and the bifractional Brownian motion.

Proposition 5.5. *Let $x \in \mathbb{R}$. The stochastic process $(u(t,x), t \geq 0)$ defined by (5.7) coincides in distribution, modulo a constant, to the bifractional Brownian motion with Hurst parameters $H = K = \frac{1}{2}$. More precisely,*

$$(u(t,x), t \geq 0) \equiv^{(d)} \left(\pi^{-\frac{1}{4}} B_t^{\frac{1}{2}, \frac{1}{2}}, t \geq 0\right) \tag{5.18}$$

where $\left(B_t^{\frac{1}{2}, \frac{1}{2}}, t \geq 0\right)$ is a bifractional Brownian motion with Hurst parameters $H = K = \frac{1}{2}$.

Proof. It suffices to compare the covariance in (5.17) and in (2.20). Indeed,

$$\mathbf{E}\pi^{-\frac{1}{4}} B_t^{\frac{1}{2}, \frac{1}{2}} \pi^{-\frac{1}{4}} B_s^{\frac{1}{2}, \frac{1}{2}} = \frac{1}{\sqrt{2\pi}}\left(\sqrt{t+s} - \sqrt{|t-s|}\right).$$

$\qquad\square$

From Proposition 5.5, we deduce immediately some properties of the
solution $(u(t,x), t \geq 0)$ given by (5.7).

Proposition 5.6. *Let u be given by (5.7). Then*

(1) For every $x \in \mathbb{R}$, the process $(u(t,x), t \geq 0)$ is $\frac{1}{4}$-self-similar.

(2) *There exists two constants* $0 < C_1 < C_2$ *such that for every* $s, t \geq 0, x \in \mathbb{R}$,

$$C_1 |t - s|^{\frac{1}{2}} \leq \mathbf{E} \, |u(t, x) - u(s, x)|^2 \leq C_2 |t - s|^{\frac{1}{2}}. \tag{5.19}$$

The constants C_1, C_2 *do not depend on* $x \in \mathbb{R}$. *In particular, for every* $x \in \mathbb{R}$, *the trajectory* $t \to u(t, x)$ *is Hölder continuous of order* δ *for every* $\delta \in \left(0, \frac{1}{4}\right)$.

(3) *We have the equality in distribution,*

$$\left(\pi^{\frac{1}{4}} u(t, x) + X_t, t \geq 0\right) \equiv^{(d)} \left(2^{\frac{1}{4}} B_t^{\frac{1}{4}}, t \geq 0\right) \tag{5.20}$$

where $(X_t, t \geq 0)$ *is Gaussian process with absolutely continuous and* C^∞ *sample paths independent by* u *and* $(B_t^{\frac{1}{4}}, t \geq 0)$ *is a fBm with Hurst index* $\frac{1}{4}$.

Proof. The self-similarity follows from Proposition 2.10 while points 2. and 3. are the consequence of Proposition 2.11 and Theorem 2.2, respectively. □

Remark 5.3. From the lower bound in (5.19) we deduce that the Hölder index is optimal. That is, the mapping $t \to u(t, x)$ cannot be Hölder continuous of order δ if $\delta > \frac{1}{4}$. Indeed, fix $x \in \mathbb{R}$ and suppose that this application is Hölder of order δ for some $\delta > \frac{1}{4}$. Then for any compact $K \subset \mathbb{R}_+$, there exists $0 < C(\omega) < \infty$ such that

$$\sup_{s, t \in K, s \neq t} \frac{|u(t, x) - u(s, x)|}{|t - s|^\delta} < C(\omega).$$

Then, by Theorem 3.2 in [Adler (1990)], we have

$$\mathbf{E} \left(\sup_{s, t \in K, s \neq t} \left| \frac{u(t, x) - u(s, x)}{|t - s|^\delta} \right|^2 \right) < \infty.$$

This implies that there exists a finite constant $L > 0$ such that

$$\mathbf{E} |u(t, x) - u(s, x)|^2 \leq L |t - s|^{2\delta}$$

which is in contradiction with the lower bound in (5.19).

Let us see the behavior of the mean square of the small increments on the solution.

Proposition 5.7. *Let* $t > 0$ *and* $x \in \mathbb{R}$ *be fixed. Then, as* $h \to 0$,

$$\frac{1}{\sqrt{h}} \mathbf{E} \, |u(t + h, x) - u(t, x)|^2 \to \frac{1}{\sqrt{\pi}}.$$

Proof. For $t > 0, x \in \mathbb{R}$ and $h > 0$, by formula (5.12),

$$\mathbf{E}\,|u(t+h,x) - u(t,x)|^2 = \frac{1}{\sqrt{2\pi}}\left(\sqrt{2t+2h} + \sqrt{2t} - 2\left(\sqrt{2t+h} - \sqrt{h}\right)\right)$$

$$= \frac{1}{\sqrt{2\pi}}\sqrt{h}\left[\sqrt{\frac{2t}{h}+2} + \sqrt{\frac{2t}{h}} - 2\left(\sqrt{\frac{2t}{h}+1} - 1\right)\right].$$

It suffices to see that the function $f(x) = \sqrt{x+2} + \sqrt{x} - 2\sqrt{x+1}$ behaves as $-\frac{1}{2}x^{-\frac{3}{2}}$ as $x \to \infty$. □

There is an alternative way to give a decomposition of the process $(u(t,x), t \geq 0)$ into a fractional Brownian motion and a stochastic process with smooth sample paths. Let $\tilde{W} = \left(\tilde{W}(t,x), t \in \mathbb{R}, x \in \mathbb{R}\right)$ be defined in the following way

$$\tilde{W}(t,x) = \begin{cases} W(t,x), & \text{if } t \geq 0, x \in \mathbb{R} \\ W^{(1)}(-t,x), & \text{if } t < 0, x \in \mathbb{R} \end{cases} \tag{5.21}$$

where W is given by (5.4) (recall that the space-time white noise can be identifies with an anisotropic Brownian sheet, see Section 3.2.1) and $W^{(1)}$ is an independent copy of W. Then $(\tilde{W}(t,x), t, x \in \mathbb{R})$ is a Brownian sheet. Define, for $t \geq 0, x \in \mathbb{R}$,

$$U(t,x)$$
$$= \int_{\mathbb{R}}\int_{\mathbb{R}} \left(G((t-u)_+, x-y) - G((-u)_+, x-y)\right)\tilde{W}(du,dy) \tag{5.22}$$

with $x_+ = \max(x,0)$, or, in other words,

$$U(t,x) = \int_{-\infty}^0 \left(G(t-u, x-y) - G(-u, x-y)\right)\tilde{W}(du,dy)$$
$$+ \int_0^t \int_{\mathbb{R}} G(t-u, x-y)W(du,dy) \tag{5.23}$$

This random field $(U(t,x), t \geq 0, x \in \mathbb{R})$ is usually named "the pinned string process" and it satisfies the following property.

Proposition 5.8. *Let $(U(t,x), t \geq 0, x \in \mathbb{R})$ be defined by (5.22). Then for every $x \in \mathbb{R}$, the centered Gaussian process $(U(t,x), t \geq 0)$ is $\frac{1}{4}$-self-similar and it has stationary increments. Actually,*

$$(U(t,x), t \geq 0) \equiv^{(d)} (C_1 B_t^{\frac{1}{4}}, t \geq 0) \tag{5.24}$$

where $B^{\frac{1}{4}}$ is a standard fBm with Hurst index $\frac{1}{4}$ and

$$C_1^2 = \frac{1}{2\pi}\int_{\mathbb{R}}\left(e^{-\frac{|\xi|^2}{2}} - 1\right)^2 |\xi|^{-2}dx + \frac{1}{\sqrt{\pi}} = \frac{\sqrt{2}}{\sqrt{\pi}}. \tag{5.25}$$

Proof. For every $0 \leq s \leq t$ and for any $c > 0$, we have by (5.22)

$$\mathbf{E}U(ct, x)U(cs, x)$$

$$= \int_{\mathbb{R}} \int_{\mathbb{R}} (G((ct - u)_+, x - y) - G((-u)_+, x - y))$$
$$\times (G((cs - u)_+, x - y) - G((-u)_+, x - y)) \, dy \, du$$

$$= c \int_{\mathbb{R}} \int_{\mathbb{R}} (G((ct - cu)_+, x - y) - G((-cu)_+, x - y))$$
$$\times (G((cs - cu)_+, x - y) - G((-cu)_+, x - y)) \, dy \, du$$

$$= c \int_{\mathbb{R}} \int_{\mathbb{R}} (G((ct - cu)_+, y) - G((-cu)_+, y))$$
$$\times (G((cs - cu)_+, y) - G((-cu)_+, y)) \, dy \, du$$

where we made the change of variables $u = c\tilde{u}$. Using the scaling property of the Green kernel (5.5),

$$G(ct, x) = \frac{1}{\sqrt{c}} G\left(t, \frac{x}{\sqrt{c}}\right), \tag{5.26}$$

we obtain, for $0 \leq s \leq t$, with $\tilde{y} = \frac{y}{\sqrt{c}}$,

$$\mathbf{E}U(ct, x)U(cs, x)$$

$$= \int_{\mathbb{R}} \int_{\mathbb{R}} \left(G\left((t - u)_+, \frac{y}{\sqrt{c}}\right) - G\left((-u)_+, \frac{y}{\sqrt{c}}\right)\right)$$
$$\times \left(G\left((s - u)_+, \frac{y}{\sqrt{c}}\right) - G\left((-u)_+, \frac{y}{\sqrt{c}}\right)\right) dy \, du$$

$$= \sqrt{c} \int_{\mathbb{R}} \int_{\mathbb{R}} (G((t - u)_+, y) - G((-u)_+, y))$$
$$\times (G((s - u)_+, y) - G((-u)_+, y)) \, dy \, du$$

$$= \mathbf{E}c^{\frac{1}{4}}U(t, x)c^{\frac{1}{4}}U(s, x)$$

and this proves the $\frac{1}{4}$-self-similarity in time of the process U. Let us prove the stationarity of its increments. By (5.23), we have for $t \geq 0, h > 0$

$$U(t + h, x) - U(h, x)$$

$$= \int_{\mathbb{R}} \int_{\mathbb{R}} (G((t + h - u)_+, x - y) - G((h - u)_+, x - y)) \, \tilde{W}(du, dy)$$

and then for $s, t \geq 0, h > 0$

$$\mathbf{E}(U(t + h, x) - U(h, x))(U(s + h, x) - U(h, x))$$

$$= \int_{\mathbb{R}} \int_{\mathbb{R}} (G((t + h - u)_+, x - y) - G((h - u)_+, x - y))$$
$$\times (G((s + h - u)_+, x - y) - G((h - u)_+, x - y)) \, dy \, du$$

$$= \int_{\mathbb{R}} \int_{\mathbb{R}} (G((t - u)_+, x - y) - G((-u)_+, x - y))$$
$$\times (G((s - u)_+, x - y) - G((-u)_+, x - y)) \, dy \, du$$

where we used $u - h = \tilde{u}$. So

$$\mathbf{E}(U(t+h,x) - U(h,x))(U(s+h,x) - U(h,x))$$
$$= \mathbf{E}(U(t,x) - U(0,x))(U(s,x) - U(0,x))$$

and this means that for every $h > 0$, the processes

$$(U(t+h,x) - U(h,x), t \geq 0) \text{ and } (U(t,x) - U(0,x), t \geq 0)$$

have the same finite dimensional distributions. By Theorem 1.2, $(U(t,x), t \geq 0)$ is modulo a constant, a fBm with Hurst parameter $\frac{1}{4}$. To get the explicit form of the constant, we need to calculate $\mathbf{E}U(1,x)^2$. We have by (5.23),

$$\mathbf{E}U(1,x)^2 = \int_{-\infty}^0 \int_{\mathbb{R}} (G(1-u, x-y) - G(-u, x-y))^2 \, dy du$$
$$+ \int_0^1 \int_{\mathbb{R}} G(1-u, x-y)^2 dy du$$

and by the identity (5.10)

$$EU(1,x)^2 = (2\pi)^{-1} \int_{-\infty}^0 \int_{\mathbb{R}} \left(e^{-\frac{(1-u)|\xi|^2}{2}} - e^{-\frac{(-u)|\xi|^2}{2}} \right)^2 d\xi du$$
$$+ (2\pi)^{-1} \int_0^1 \int_{\mathbb{R}} e^{-(1-u)|\xi|^2} d\xi du$$
$$= (2\pi)^{-1} \int_0^\infty du \left(\int_{\mathbb{R}} e^{-u|\xi|^2} \left(e^{-\frac{|\xi|^2}{2}} - 1 \right)^2 d\xi \right)$$
$$+ (2\pi)^{-1} \int_0^1 du (\sqrt{2u})^{-1} \int_{\mathbb{R}} e^{-\frac{|\xi|^2}{2}} d\xi$$
$$= (2\pi)^{-1} \int_{\mathbb{R}} \left(e^{-\frac{|\xi|^2}{2}} - 1 \right)^2 |\xi|^{-2} d\xi + \frac{1}{\sqrt{\pi}}$$

and by using the integration by parts,

$$(2\pi)^{-1} \int_{\mathbb{R}} \left(e^{-\frac{|\xi|^2}{2}} - 1 \right)^2 |\xi|^{-2} d\xi = \frac{\sqrt{2}}{\sqrt{\pi}} - \frac{1}{\sqrt{\pi}}.$$

Then

$$\mathbf{E}U(1,x)^2 = \frac{\sqrt{2}}{\sqrt{\pi}}$$

and this is equal with C_1^2 from (5.25). $\qquad\square$

Let us now set, for $t \geq 0, x \in \mathbb{R}$,

$$Y(t, x) = u(t, x) - U(t, x). \tag{5.27}$$

Then we have the following result.

Proposition 5.9. *Let Y be given by (5.27). Then $t \to Y(t, x)$ is absolutely continuous is of class C^∞ on $(0, \infty)$.*

Proof. The first remark is that the random field given by (5.27) is well-defined. By (5.23), for every $t > 0, x \in \mathbb{R}$,

$$Y(t, x) = - \int_{-\infty}^{0} \int_{\mathbb{R}} (G(t - u, x - y) - G(-u, x - y)) \, \tilde{W}(du, dy).$$

Set, for every $t > 0$ and $x \in \mathbb{R}$,

$$Y'(t, x) = - \int_{-\infty}^{0} \int_{\mathbb{R}} \frac{\partial}{\partial t} G(t - u, x - y) \tilde{W}(du, dy) \tag{5.28}$$

the formal derivative of Y with respect to the time variable t. We will have

$$
\begin{aligned}
\mathbf{E}|Y'(t, x)|^2 &= \int_{-\infty}^{0} \int_{\mathbb{R}} \left| \frac{\partial}{\partial t} G(t - u, x - y) \right|^2 dy ds \\
&= (2\pi)^{-1} \int_{-\infty}^{0} \int_{\mathbb{R}} \left| \mathcal{F} \frac{\partial}{\partial t} G(t - u, \cdot)(\xi) \right|^2 d\xi du \\
&= (2\pi)^{-1} \int_{-\infty}^{0} \int_{\mathbb{R}} \left| \frac{\partial}{\partial t} e^{-\frac{(t-u)|\xi|^2}{2}} \right|^2 d\xi du \\
&= \frac{1}{4} (2\pi)^{-1} \int_{-\infty}^{0} \int_{\mathbb{R}} \left(e^{-\frac{(t-u)|\xi|^2}{2}} |\xi|^2 \right)^2 d\xi du
\end{aligned}
$$

and thus

$$\mathbf{E}|Y'(t, x)|^2 = \frac{1}{4} (2\pi)^{-1} \int_{-\infty}^{0} ds(t - s)^{-\frac{5}{2}} \int_{\mathbb{R}} d\xi |\xi|^4 e^{-|\xi|^2} < \infty.$$

This shows that the mapping $t \to Y(t, x) : (0, \infty) \to \mathbb{R}$ is absolute continuous and differentiable with respect to Y and its derivative $\frac{d}{dt} Y(t, x)$ is given by (5.28). The same argument holds to get the nth times differentiability of $Y(t, x)$ with respect to t. Let

$$Y^{(n)}(t, x) = - \int_{-\infty}^{0} \int_{\mathbb{R}} \frac{\partial^n}{\partial t^n} G(t - u, x - y) W(du, dy)$$

for $t \geq 0, x \in \mathbb{R}$. Its $L^2(\Omega)$-norm can be computed as follows

$$\mathbf{E}\left|Y^{(n)}(t,x)\right|^2 = \int_{-\infty}^0 \int_{\mathbb{R}} \left|\frac{\partial^n}{\partial t^n} G(t - u, x - y)\right|^2 dy\,du$$

$$= (2\pi)^{-1} \int_{-\infty}^0 \int_{\mathbb{R}} \left|\frac{\partial^n}{\partial t^n} e^{-\frac{(t-u)|\xi|^2}{2}}\right|^2 d\xi\,du$$

$$= (2\pi)^{-1} 2^{-2n} \int_{-\infty}^0 \int_{\mathbb{R}} |\xi|^{2n} e^{-(t-u)|\xi|^2} d\xi\,du$$

and consequently

$$\mathbf{E}\left|Y^{(n)}(t,x)\right|^2 = (2\pi)^{-1} 2^{-2n} \int_{-\infty}^0 du(t-u)^{-\frac{1}{2}-n} \int_{\mathbb{R}} d\xi |\xi|^{2n} e^{-|\xi|^2} < \infty.$$

This shows that $t \to Y(t,x)$ is of class C^∞ on $(0, \infty)$. $\qquad\square$

Let us put together the previous results in Propositions 5.8 and 5.9.

Theorem 5.1. *Let $(u(t,x), t \geq 0, x \in \mathbb{R})$ be defined by (5.7) and assume $x \in \mathbb{R}$ is fixed. Then*

$$(u(t,x), t \geq 0) \equiv^{(d)} \left(\left(\frac{2}{\pi}\right)^{-\frac{1}{4}} B_t^{\frac{1}{4}} + Y(t,x), t \geq 0\right) \tag{5.29}$$

where $(B_t^{\frac{1}{4}}, t \geq 0)$ is a fBm with $H = \frac{1}{4}$ and Y is such that $t \to Y(t,x)$ is absolute continuous and of class C^∞ on $(0, \infty)$. Moreover, for every $t > 0$, $(u(t,x), x \in \mathbb{R})$ and $(Y(t,x), x \in \mathbb{R})$ are independent Gaussian stochastic processes.

Proof. It suffices to apply Propositions 5.8 and 5.9. $\qquad\square$

Remark 5.4.

In formula (5.29) we retrieve the result stated in (5.20). On the other hand, in Theorem 5.1 we have the explicit form of the regular process Y as a integral with respect to the white noise.

5.1.4 *Behavior in space*

Let us analyze the distribution of the process $(u(t,x), x \in \mathbb{R})$ given by (5.7) with $t \geq 0$ fixed. This is also a centered Gaussian process but with rather different properties than its temporal counterpart studied in the previous paragraph. Its covariance can directly be obtained from (5.12).

Proposition 5.10. *For every $t \geq 0$ and $x, y \in \mathbb{R}$, we have*

$$\mathbf{E}u(t,x)u(t,y) = \frac{1}{\sqrt{2\pi}} \left(\sqrt{2t}e^{-\frac{|x-y|^2}{4t}} - |x-y| + \sqrt{2}|x-y|erf\left(\frac{|x-y|}{\sqrt{4t}}\right) \right)$$
(5.30)

with the error function erf given by (5.13).

Proof. Take $s = t$ in (5.12) and observe that $erf(\infty) = \frac{1}{\sqrt{2}}$. $\square$

From the above formula we get immediately the spatial stationarity of the solution (5.7).

Corollary 5.1. *For every $t \geq 0$, the Gaussian process $(u(t,x), x \in \mathbb{R})$ is stationary.*

Proof. From (5.30) we see that the spatial covariance of the Gaussian process $(u(t,x), x \in \mathbb{R})$ is a function of $|x - y|$. We then use Proposition 1.2. $\square$

The expression (5.30) of the covariance does not give directly the relationship between the spatial process $(u(t,x), x \in \mathbb{R})$ and the fractional or bifractional Brownian motion. On the other hand, we will see that such a relationships exists. We will have a decomposition as in Theorem 5.1 into a (two-sided) Wiener process and a Gaussian process with very smooth sample paths. Let us start by defining and analyzing the regular part of this decomposition.

For $t > 0$ and $x \in \mathbb{R}$, we denote

$$S(t,x) = \int_t^\infty \int_{\mathbb{R}} (G(s,y) - G(s,x-y)) W(ds,dy). \tag{5.31}$$

Proposition 5.11. *Let S be given by (5.31). Then $(S(t,x), t > 0, x \in \mathbb{R})$ is a well-defined centered Gaussian random field and it satisfies the following properties:*

(1) For every $t > 0, x, y \in \mathbb{R}$

$$\mathbf{E}\,|S(t,x) - S(t,y)|^2 \leq C|x-y|^2 \tag{5.32}$$

with $C = C_t = \frac{1}{\sqrt{\pi t}} > 0$.

(2) For $t > 0$ fixed, the mapping $x \to S(t,x)$ is absolutely continuous and it has C^∞-sample paths.

Proof. Let us show first that

$$\mathbf{E}S(t,x)^2 < \infty$$

for every $t > 0, x \in \mathbb{R}$. We have, by Parseval's identity (5.10),

$$\mathbf{E}|S(t,x)|^2 = \int_t^\infty \int_{\mathbb{R}} |G(s,y) - G(s, x-y)|^2 \, dy \, ds$$

$$= \frac{1}{2\pi} \int_t^\infty \int_{\mathbb{R}} |\mathcal{F}(G(s,\cdot) - G(s, x - \cdot))(\xi)|^2 \, d\xi \, ds$$

and since by (5.11), for $s > 0, x, \xi \in \mathbb{R}$

$$\mathcal{F}G(s, x - \cdot)(\xi) = \int_{\mathbb{R}} dy \, e^{i\xi y} G(s, x-y) = e^{i\xi x} \overline{\mathcal{F}G(s,\cdot)\xi}$$

$$= e^{ix\xi} e^{-\frac{s|\xi|^2}{2}},$$

(see (5.11)), we can write

$$\mathbf{E}|S(t,x)|^2 = \frac{1}{2\pi} \int_t^\infty \int_{\mathbb{R}} e^{-s|\xi|^2} \left|1 - e^{i\xi x}\right|^2 d\xi \, ds$$

$$= \frac{1}{\pi} \int_t^\infty \int_{\mathbb{R}} e^{-s|\xi|^2} (1 - \cos(x\xi)) \, d\xi \, ds,$$

where we used the simple fact that

$$\left|e^{ix} - 1\right|^2 = 2(1 - \cos(x)), \quad x \in \mathbb{R}.$$

Next, by calculating the integral ds we get

$$\mathbf{E}|S(t,x)|^2 = \frac{1}{\pi} \int_{\mathbb{R}} d\xi \frac{1 - \cos(x\xi)}{|\xi|^2} e^{-t|\xi|^2}$$

$$= \frac{2}{\pi} \int_0^\infty d\xi \frac{1 - \cos(x\xi)}{|\xi|^2} e^{-t|\xi|^2}.$$

The above integral is finite because, via the bound $|1 - \cos(x\xi)| \le C|\xi|^2|x|^2$ for $x, \xi \in \mathbb{R}$,

$$\int_0^\infty d\xi \frac{1 - \cos(x\xi)}{|\xi|^2} e^{-t|\xi|^2}$$

$$\le |x|^2 \int_0^\infty d\xi \, e^{-t|\xi|^2} = \frac{\sqrt{\pi}}{2\sqrt{t}} |x|^2 < \infty.$$

So $(S(t,x), t > 0, x \in \mathbb{R})$ is a well-defined Gaussian random field with zero expectation.

Let us now estimate $\mathbf{E}\,|S(t,x_2)-S(t,x_1)|^2$ for $x_1,x_2\in\mathbb{R}$ and $t>0$. By applying again Parseval's relation and by using the Fourier transform of the Green kernel (5.11)

$$
\begin{aligned}
\mathbf{E}\,|S(t,x_2)-S(t,x_1)|^2 &= \int_t^\infty \int_{\mathbb{R}} (G(s,x_1-y)-G(s,x_2-y))^2\,dyds\\
&= \frac{1}{2\pi}\int_t^\infty \int_{\mathbb{R}} \left|e^{-ix_2\xi}-e^{ix_1\xi}\right|^2 e^{-s|\xi|^2}\,d\xi ds\\
&= \frac{1}{2\pi}\int_t^\infty \int_{\mathbb{R}} \left|e^{-i(x_1-x_2)\xi}-1\right|^2 e^{-s|\xi|^2}\,d\xi ds\\
&= \frac{1}{\pi}\int_t^\infty \int_{\mathbb{R}} (1-\cos(x_2-x_1)\xi)e^{-s|\xi|^2}\,d\xi ds.
\end{aligned}
$$

Next, by calculating the integral ds and using $|1-\cos(x)|\le C|x|^2$ for $x\in\mathbb{R}$, we obtain

$$
\begin{aligned}
\mathbf{E}\,|S(t,x_2)-S(t,x_1)|^2 &= \frac{1}{\pi}\int_{\mathbb{R}} d\xi e^{-t|\xi|^2}\frac{1-\cos((x_1-x_2)\xi)}{|\xi|^2}\\
&= \frac{2}{\pi}\int_0^\infty d\xi e^{-t|\xi|^2}\frac{1-\cos((x_1-x_2)\xi)}{|\xi|^2}\\
&\le \frac{2}{\pi}|x_1-x_2|^2 \int_0^\infty d\xi e^{-t|\xi|^2}\\
&= C_t|x_1-x_2|^2
\end{aligned}
$$

with $C_t=\frac{2}{\pi}\int_0^\infty d\xi e^{-t|\xi|^2}<\infty$.

Fix $t>0$ (notice the strict inequality!) and let us show that the process $(S(t,x),x\in\mathbb{R})$ is absolutely continuous and it has C^∞-sample paths. Let

$$
\frac{dS}{dx}(t,x)=\int_t^\infty \int_{\mathbb{R}} -\frac{dG}{dx}(s,x-y)W(ds,dy), \tag{5.33}
$$

the formal derivative of S with respect to the variable $x\in\mathbb{R}$. We have

$$
\frac{dS}{dx}(t,x)=\int_t^\infty \int_{\mathbb{R}} \frac{1}{\sqrt{2\pi s}}e^{-\frac{(x-y)^2}{2s}}\frac{x-y}{s}W(ds,dy)
$$

and

$$
\begin{aligned}
\mathbf{E}\left|\frac{dS}{dx}(t,x)\right|^2 &= \frac{1}{2\pi}\int_t^\infty s^{-1}\int_{\mathbb{R}} e^{-\frac{(x-y)^2}{s}}\frac{(x-y)^2}{s^2}\,dyds\\
&= \frac{1}{2\pi}\int_t^\infty s^{-1}\int_{\mathbb{R}} e^{-\frac{y^2}{s}}\frac{y^2}{s^2}\,dyds
\end{aligned}
$$

and with the change of variable $z = \frac{y}{\sqrt{s}}$ in the integral dy,

$$\mathbf{E}\left|\frac{dS}{dx}(t,x)\right|^2 = \frac{1}{2\pi}\int_t^\infty s^{-\frac{3}{2}}\int_\mathbb{R} e^{-z^2}z^2 dz < \infty \text{ for every } t > 0.$$

This implies that the path $x \to S(t,x)$ is absolutely continuous and $\frac{dS}{dx}(t,x)$ is given by (5.33). Similarly, for every $n \geq 1$, set

$$\frac{d^n S}{dx^n}(t,x) = \int_t^\infty \int_\mathbb{R} \frac{d^n}{dx^n}\left(G(s,y) - G(s,x-y)\right)W(ds,dy).$$

We will have

$$\mathbf{E}\left|\frac{d^n S}{dx^n}(t,x)\right|^2 = C\int_t^\infty \int_\mathbb{R} s^{-1}e^{-\frac{y^2}{s}}\frac{y^{2n}}{s^{2n}}dyds$$

$$= C\int_t^\infty s^{-n-\frac{1}{2}}ds < \infty$$

and this implies the n-th differentiability of $S(t,x)$ with respect to the spatial variable $x \in \mathbb{R}$. $\square$

Let us denote, for $t > 0, x \in \mathbb{R}$,

$$V(t,x) = u(t,x) - S(t,x). \tag{5.34}$$

We will show that $(V(t,x), x \in \mathbb{R})$ is (modulo a constant) a Brownian motion for $t > 0$ fixed.

Proposition 5.12. *Let V be given by (5.34). Then for fixed $t > 0$, the process $(U(t,x), x \in \mathbb{R})$ has the same law as*

$$\left(\sqrt{C_0}B(x), x \in \mathbb{R}\right)$$

with $(B(x), x \in \mathbb{R})$ a standard (two-sided) Brownian motion and with

$$C_0 = \frac{2}{\pi}\int_0^\infty \frac{1 - \cos(|\xi|)}{|\xi|^2}d\xi < \infty. \tag{5.35}$$

Proof. Clearly, it is a centered Gaussian process. We will compute $\mathbf{E}(V(t,x) - V(t,y))^2$ for every $x, y \in \mathbb{R}$. We have, by (5.34),

$$V(t,x) - V(t,y) = u(t,x) - u(t,y) - (S(t,x) - S(t,y))$$

$$= \int_0^t \int_\mathbb{R}\left(G(t-s,x-z) - G(t-s,y-z)\right)W(ds,dz)$$

$$+ \int_t^\infty \int_\mathbb{R}\left(G(s,x-z) - G(s,y-z)\right)W(ds,dz).$$

Notice that for every $t > 0, x \in \mathbb{R}$,

$$\int_0^t \int_{\mathbb{R}} (G(t - s, x - z) - G(t - s, y - z)) \, W(ds, dz)$$

and

$$\int_t^{\infty} \int_{\mathbb{R}} (G(s, x - z) - G(s, y - z)) \, W(ds, dz)$$

are independent centered Gaussian random variables. Thus

$$
\begin{aligned}
\mathbf{E} \left(V(t, x) - V(t, y) \right)^2 &= \int_0^t \int_{\mathbb{R}} (G(t - s, x - z) - G(t - s, y - z))^2 \, dz ds \\
&\quad + \int_t^{\infty} \int_{\mathbb{R}} (G(s, x - z) - G(s, y - z))^2 \, dz ds \\
&= \int_0^t \int_{\mathbb{R}} (G(s, x - z) - G(s, y - z))^2 \, dz ds \\
&\quad + \int_t^{\infty} \int_{\mathbb{R}} (G(s, x - z) - G(s, y - z))^2 \, dz ds \\
&= \int_0^{\infty} \int_{\mathbb{R}} (G(s, x - z) - G(s, y - z))^2 \, dz ds.
\end{aligned}
$$

Now, by applying Parseval's relation and proceeding as above in the proof of Proposition 5.11,

$$
\begin{aligned}
\mathbf{E} \left(V(t, x) - V(t, y) \right)^2 &= \frac{1}{\pi} \int_0^{\infty} \int_{\mathbb{R}} (1 - \cos((x - y)\xi)) \, e^{-s|\xi|^2} d\xi ds \\
&= \frac{1}{\pi} \int_{\mathbb{R}} \frac{1 - \cos((x - y)\xi)}{|\xi|^2} d\xi \\
&= \frac{2}{\pi} \int_0^{\infty} \frac{1 - \cos(|\xi|)}{|\xi|^2} d\xi |x - y| \\
&= C_0 |x - y|
\end{aligned}
$$

with C_0 given by (5.35). Hence, for any $x, y \in \mathbb{R}$,

$$
\begin{aligned}
\mathbf{E} V(t, x) V(t, y) &= \frac{1}{2} \left[\mathbf{E} \left(V(t, x) - V(t, y) \right)^2 - \mathbf{E} V(t, x)^2 - \mathbf{E} V(t, y)^2 \right] \\
&= \frac{C_0}{2} \left(|x| + |y| - |x - y| \right)
\end{aligned}
$$

and this implies the conclusion. $\qquad \square$

We can conclude with a spatial decomposition theorem of the solution to the heat-equation with space-time white noise.

Theorem 5.2. *Let u be given by (5.7) and assume that $t > 0$ is fixed. Then we have*

$$(u(t,x), x \in \mathbb{R}) \equiv^{(d)} \left(\sqrt{C_0} B(x) + S(t,x), x \in \mathbb{R} \right) \qquad (5.36)$$

where $(B(x), x \in \mathbb{R})$ is a two-sided Brownian motion and S is given by (5.31). Moreover, the random fields

$$(u(t,x), x \in \mathbb{R}) \text{ and } (S(t,x), x \in \mathbb{R})$$

are independent.

Proof. It is an immediate consequence of Propositions 5.11 and 5.12. □

Remark 5.5. We can also state the result in Theorem 5.2 by saying that for every $t > 0$,

$$(u(t,x) - S(t,x), x \in \mathbb{R})$$

is a Wiener process on the whole real line. We mention again the independence of $(u(t,x), x \in \mathbb{R})$ and $(S(t,x), x \in \mathbb{R})$ for every $t > 0$.

The decomposition in (5.2) will lead to useful properties for the mild solution, concerning its trajectories or its spatial variations. We first deduce the following sharp inequality for the spatial increment of the solution.

Proposition 5.13. *Let u be defined by (5.7). Then for every $t \geq t_0 > 0$ and for $x, y \in \mathbb{R}$ with $|x - y|$ small enough, there exists $0 < C_1 < C_2$ such that*

$$C_1 |x - y| \leq \mathbf{E} \left(u(t,x) - u(t,y) \right)^2 \leq C_2 |x - y|. \qquad (5.37)$$

In particular, for every $t > 0$, the mapping $x \to u(t,x)$ is δ-Hölder continuous, for every $\delta \in \left(0, \frac{1}{2} \right)$.

Proof. The second inequality follows directly from the decomposition in Theorem 5.2. Concerning the lower bound, by the independence of u and S we have by the decomposition (5.36),

$$C_0 \mathbf{E} |B(x) - B(y)|^2 = \mathbf{E} |u(t,x) - u(t,y)|^2$$
$$+ \mathbf{E} |S(t,x) - S(t,y)|^2$$

so, by the bound (5.32)

$$\mathbf{E} |u(t,x) - u(t,x)|^2 = C_0 \mathbf{E} |B(x) - B(y)|^2 - \mathbf{E} |S(t,x) - S(t,x)|^2$$
$$\geq C_0 \mathbf{E} |B(x) - B(y)|^2 - C|x - y|^2$$
$$= C_0 |x - y| - C|x - y|^2$$
$$\geq C_1 |x - y|$$

for $|x - y|$ small enough.

The Hölder continuity follows from the inequality (easily derived from the upper bound in (5.37))

$$\mathbf{E} |u(t, x) - u(t, y)|^p \leq C_2 |x - y|^{\frac{p}{2}}$$

for all $p \geq 1, x \in \mathbb{R}$ and by the Kolmogorov continuity criterion. $\qquad\square$

Remark 5.6.

(1) We notice that the solution (5.7) is δ-Hölder continuous with $\delta \in \left(0, \frac{1}{4}\right)$ in time and $\delta \in \left(0, \frac{1}{2}\right)$ in space. That is, the spatial regularity is twice the temporal regularity.

(2) The Hölder regularity in space is sharp, due to the lower bound in (5.37). As in Remark 5.3, we can prove that the mapping $x \to u(t, x)$ cannot be δ-Hölder continuous if $\delta > \frac{1}{2}$.

We can also get the continuity of the solution (5.7) in both (temporal and spatial) variables.

Proposition 5.14. *For every $x, y \in \mathbb{R}$, $s, t \geq 0$, we have for $p \geq 2$,*

$$\mathbf{E} |u(t, x) - u(s, y)|^p \leq C_p \left[|t - s|^{\frac{p}{4}} + |x - y|^{\frac{p}{2}} \right].$$

Consequently the trajectories $(t, x) \to u(t, x)$ are almost surely continuous on $(0, \infty) \times \mathbb{R}$.

Proof. Let $s, t \geq 0, x, y \in \mathbb{R}$. Since $u(t, x) - u(s, y)$ is a Gaussian random variable (below $C_p > 0$ may change from one place to another)

$$\mathbf{E} |u(t, x) - u(s, y)|^p = C_p \left(\mathbf{E} |u(t, x) - u(s, y)|^2 \right)^{\frac{p}{2}}$$

$$\leq C_p \left(\mathbf{E} |u(t, x) - u(s, x)|^2 + \mathbf{E} |u(s, x) - u(s, y)|^2 \right)^{\frac{p}{2}}$$

and by using Proposition 5.6, point 2. and Proposition 5.13, we get

$$\mathbf{E} |u(t, x) - u(s, y)|^p \leq C_p \left(|t - s|^{\frac{1}{2}} + |x - y| \right)^{\frac{p}{2}}$$

$$\leq C_p \left[|t - s|^{\frac{p}{4}} + |x - y|^{\frac{p}{2}} \right].$$

The existence of a bicontinuous version in (t, x) follows from Theorem 3.2.

$\qquad\square$

5.2 The fractional stochastic heat equation with space-time white noise

In this part, we consider the stochastic heat equation, still driven by the space-time white noise, but with the fractional Laplacian operator instead of the standard Laplacian. As mentioned above, this equation modelizes the heat flow under a random perturbation in a non-homogenous medium.

5.2.1 *General properties*

Consider the stochastic partial differential equation

$$\frac{\partial}{\partial t}u(t,x) = -(-\Delta)^{\frac{\alpha}{2}}u(t,x) + \dot{W}(t,x), \qquad t \geq 0, x \in \mathbb{R}^d. \tag{5.38}$$

with vanishing initial condition

$$u(0,x) = 0 \text{ for every } x \in \mathbb{R}^d.$$

In (5.38), $-(-\Delta)^{\frac{\alpha}{2}}$ denotes the fractional Laplacian with exponent $\frac{\alpha}{2}$, $\alpha \in (1,2]$ and W is space-time white noise, i.e. $(W(t,A), t \geq 0, A \in \mathcal{B}(\mathbb{R}^d))$ is a centered Gaussian field with covariance given by (5.4).

We refer to [Chen and Dalang (2015)], [Debbi and Dozzi (2005)], [Jacob and Leopold (1993)], [Jacob *et al.* (2015)], [Jiang *et al.* (2010)] for the precise definition and other properties of the fractional Laplacian operator. We will here mainly use the expression of the Green kernel G_α (or the fundamental solution) associated to the fractional Laplacian, i.e. the deterministic kernel that solves the heat equation without noise

$$\frac{\partial}{\partial t}u(t,x) = -(-\Delta)^{\frac{\alpha}{2}}u(t,x).$$

This Green kernel is defined through its Fourier transform

$$\mathcal{F}G_\alpha(t,\cdot)(\xi) = e^{-t\|\xi\|^\alpha}, \qquad t > 0, \xi \in \mathbb{R}^d \tag{5.39}$$

where $\mathcal{F}G_\alpha(t,\cdot)$ is the Fourier transform of the function $y \to G_\alpha(t,y)$. This means that

$$G_\alpha(t,x) = \int_{\mathbb{R}^d} e^{-i\langle\xi,x\rangle} e^{-t|\xi|^\alpha} d\xi, \quad t > 0, x \in \mathbb{R}^d.$$

For $\alpha = 2$, we have

$$G_\alpha(t,x) = G(2t,x), t > 0, x \in \mathbb{R}^d \tag{5.40}$$

with G the Green kernel given by (5.5). We did not find exactly $G(t,x)$ for $\alpha = 2$, this is due to the constant $\frac{1}{2}$ which appears in front of the Laplacian in (5.3) and it does not appear in (5.38). Other important properties of the fractional Laplacian operator are listed below, see among others [Chen and Dalang (2015)], [Debbi and Dozzi (2005)].

(1) For every $t > 0$, $G_\alpha(t, \cdot)$ is a density function (actually it is the density of a d-dimensional Lévy stable process at time t). In particular, we have

$$\int_{\mathbb{R}^d} G_\alpha(t, x) dx = 1. \tag{5.41}$$

(2) For every $t > 0$, the kernel $G_\alpha(t, x)$ is real valued, positive, and symmetric in x.

(3) The operator G_α satisfies the semigroup property, i.e.

$$G_\alpha(t + s, x) = \int_{\mathbb{R}^d} G_\alpha(t, z) G_\alpha(s, x - z) dz \tag{5.42}$$

for $0 < s \leq t$ and $x \in \mathbb{R}^d$.

(4) G_α is infinitely differentiable with respect to x, with all the derivatives bounded and converging to zero as $\|x\| \to \infty$. Moreover, we have the scaling property (which also holds for $\alpha = 2$)

$$G_\alpha(t, x) = t^{-\frac{d}{\alpha}} G_\alpha(1, t^{-\frac{1}{\alpha}} x). \tag{5.43}$$

For $d = 1$ and $\alpha = 2$, the formula (5.43) coincides with (5.26).

(5) There exist two constants $0 < K'_\alpha < K_\alpha$ such that

$$K'_\alpha \frac{1}{(1 + \|x\|)^{d+\alpha}} \leq |G_\alpha(1, x)| \leq K_\alpha \frac{1}{(1 + \|x\|)^{d+\alpha}} \tag{5.44}$$

for all $x \in \mathbb{R}^d$. Together with the scaling property, this further translates into

$$K'_\alpha \frac{t^{-\frac{d}{\alpha}}}{\left(1 + \|t^{-\frac{1}{\alpha}} x\|\right)^{d+\alpha}} \leq |G_\alpha(t, x)| \leq K_\alpha \frac{t^{-\frac{d}{\alpha}}}{\left(1 + \|t^{-\frac{1}{\alpha}} x\|\right)^{d+\alpha}}. \tag{5.45}$$

The mild solution to (5.38) is understood in the mild sense, i.e.

$$u(t, x) = \int_0^t \int_{\mathbb{R}^d} G_\alpha(t - u, x - z) W(du, dz) \tag{5.46}$$

where the above integral $W(du, dz)$ is a Wiener integral with respect to the Gaussian noise W.

First, we notice that the solution exists only in spatial dimension $d = 1$.

Proposition 5.15. *Let* $\big(u(t, x), t \geq 0, x \in \mathbb{R}^d\big)$ *be given by (5.46). Then the solution is well-defined if and only if* $d = 1$. *Moreover, in this case, for every* $T > 0$ *and for every* $p \geq 1$

$$\sup_{t \in [0, T], x \in \mathbb{R}} \mathbb{E} |u(t, x)|^p < \infty. \tag{5.47}$$

Proof. From the Wiener isometry, the Parseval identity (5.10) and the expression of the Fourier transform (5.39), we have for every $t > 0, x \in \mathbb{R}^d$

$$
\mathbf{E}u(t,x)^2 = \int_0^t du \int_{\mathbb{R}^d} dz \, |G_\alpha(t-u, x-z)|^2
$$

$$
= (2\pi)^{-d} \int_0^t du \int_{\mathbb{R}^d} d\xi \, |\mathcal{F}G_\alpha(u, \cdot)(\xi)|^2
$$

$$
= (2\pi)^{-d} \int_0^t du \int_{\mathbb{R}^d} d\xi e^{-2u\|\xi\|^\alpha} = c_{d,\alpha} \int_0^t du \, u^{-\frac{d}{\alpha}}
$$

with $c_{d,\alpha} = (2\pi)^{-d} \int_{\mathbb{R}^d} d\xi e^{-2\|\xi\|^\alpha} < \infty$. The integral $\int_0^t u^{-\frac{d}{\alpha}} du$ is finite if and only if

$$
1 - \frac{d}{\alpha} > 0
$$

which means $d < \alpha$ or equivalently $d = 1$ since $\alpha \in (1, 2]$. Moreover, for every $t > 0, x \in \mathbb{R}$,

$$
\mathbf{E}u(t,x)^2 = c_{1,\alpha} \frac{1}{1 - \frac{1}{\alpha}} t^{1-\frac{1}{\alpha}}
$$

which implies (5.47), since $u(t, x)$ is a Gaussian random variables. $\square$

A more explicit expression of the constant $c_{1,\alpha}$ is given in Proposition 5.16. Next, we will focus on the probability distribution of the solution in spatial dimension $d = 1$. We will treat separately the behavior in time and in space.

Remark 5.7.

(1) The necessary and sufficient condition for the existence of the solution (i.e. $d = 1$) can be also obtained via the double inequality (5.45).
(2) We can show that, if $d = 1$, the mild solution (5.46) is also a solution to the heat equation in the weak sense (see Definition 5.1 and the proof of Proposition 5.3).
(3) For $\alpha = 2$, we have

$$
v(t,x) = u\left(\frac{t}{2}, x\right), \quad t \geq 0, x \in \mathbb{R}
$$

where v is the mild solution to the standard heat equation defined by (5.7). This comes from (5.40).

5.2.2 Behavior in time

We consider here the process $(u(t, x), t \geq 0)$ with $x \in \mathbb{R}$ fixed. The distribution and the properties of this Gaussian process will follow easily from the computation of its covariance.

Proposition 5.16. *For every $s, t \geq 0$ and $x \in \mathbb{R}$ we have*

$$\mathbf{E}u(t, x)u(s, x) = c_{1,\alpha} \left[(t + s)^{1-\frac{1}{\alpha}} - |t - s|^{1-\frac{1}{\alpha}} \right] \tag{5.48}$$

where

$$c_{1,\alpha} = \frac{1}{2\pi(\alpha - 1)} \Gamma\left(\frac{1}{\alpha}\right).$$

Consequently, the process $(u(t, x), t \geq 0)$ has the same law as the Gaussian process

$$\left(c_{2,\alpha} B_t^{\frac{1}{2}, 1-\frac{1}{\alpha}}, \right),$$

where $(B_t^{\frac{1}{2}, 1-\frac{1}{\alpha}}, t \geq 0)$ is a bifractional Brownian motion with Hurst parameters $H = \frac{1}{2}$ and $K = 1 - \frac{1}{\alpha}$ and

$$c_{2,\alpha}^2 = c_{1,\alpha} 2^{1-\frac{1}{\alpha}}. \tag{5.49}$$

In particular, for any $t > 0, x \in \mathbb{R}$,

$$u(t, x) \sim N\left(0, c_{1,\alpha} 2^{1-\frac{1}{\alpha}} t^{1-\frac{1}{\alpha}}\right) = N\left(0, c_{2,\alpha} t^{1-\frac{1}{\alpha}}\right).$$

Proof. We follow the lines of the proof of Proposition 5.15. Assuming that $0 \leq s \leq t$, we have from (5.39)

$$\mathbf{E}u(t, x)u(s, x)$$
$$= \int_0^{t \wedge s} du \int_{\mathbb{R}} dz G_\alpha(t - u, x - z) G_\alpha(s - u, x - z)$$
$$= (2\pi)^{-1} \int_0^{t \wedge s} du \int_{\mathbb{R}} d\xi \mathcal{F}G_\alpha(t - u, \cdot)(\xi) \overline{\mathcal{F}G_\alpha(s - u, \cdot)(\xi)}$$
$$= (\pi)^{-1} \int_0^s du \int_0^\infty d\xi e^{-(t+s-2u)|\xi|^\alpha}$$
$$= (\pi)^{-1} \frac{2}{\alpha} \int_0^\infty d\xi \, |\xi|^{\frac{2}{\alpha}-1} e^{-|\xi|^2} \int_0^s du(t + s - 2u)^{-1/\alpha}$$

and we have

$$\int_0^\infty d\xi \, |\xi|^{\frac{2}{\alpha}-1} e^{-|\xi|^2} = \frac{1}{2} \Gamma\left(\frac{1}{\alpha}\right)$$

and

$$\int_0^s du (t + s - 2u)^{-1/\alpha} = \frac{1}{2} \frac{1}{1 - \frac{1}{\alpha}} \left[(t+s)^{1-\frac{1}{\alpha}} - |t-s|^{1-\frac{1}{\alpha}} \right].$$

Then

$$\mathbf{E} u(t,x) u(s,x) = \frac{1}{2} (\pi)^{-1} \frac{2}{\alpha} \frac{1}{2} \Gamma \left(\frac{1}{\alpha} \right) \frac{1}{1 - \frac{1}{\alpha}} \left[(t+s)^{1-\frac{1}{\alpha}} - |t-s|^{1-\frac{1}{\alpha}} \right]$$

$$= \frac{1}{2\pi(\alpha - 1)} \Gamma \left(\frac{1}{\alpha} \right) \left[(t+s)^{1-\frac{1}{\alpha}} - |t-s|^{1-\frac{1}{\alpha}} \right].$$

$\square$

Remark 5.8. If $\alpha = 2$, then $c_{1,\alpha} = \frac{1}{2\sqrt{\pi}}$ and $c_{2\alpha} = \sqrt{2\pi}$ since $\Gamma(\frac{1}{2}) = \sqrt{\pi}$. We retrieve (modulo a constant, coming from the appearance of $\frac{1}{2}$ in front of the Laplacian in (5.3)) the formula (5.17).

From Proposition 5.16, we can deduce many properties of the process $t \to u(t,x)$.

Proposition 5.17. *For every $x \in \mathbb{R}$,*

(1) the process $(u(t,x), t \geq 0)$ is self-similar of order $\frac{1}{2} \left(1 - \frac{1}{\alpha} \right)$

(2) the trajectory $t \to u(t,x)$ is Hölder continuous of order δ, for any $\delta \in \left(0, \frac{1}{2} \left(1 - \frac{1}{\alpha} \right) \right)$.

(3) there exist two constants (depending on α) $0 < C_1 < C_2$ such that for every $s, t \geq 0$,

$$C_1 |t-s|^{1-\frac{1}{\alpha}} \leq \mathbf{E} |u(t,x) - u(s,x)|^2 \leq C_2 |t-s|^{1-\frac{1}{\alpha}}. \qquad (5.50)$$

(4) We have the following decomposition in law:

$$(u(t,x) + Y_t, t \geq 0) \equiv^{(d)} \left((c_{2,\alpha} 2^{\frac{1}{2\alpha}}) B_t^{\frac{1}{2}(1-\frac{1}{\alpha})}, t \geq 0 \right) \qquad (5.51)$$

where Y is a Gaussian process with absolute continuous paths and

$$\left(B_t^{\frac{1}{2}(1-\frac{1}{\alpha})}, t \geq 0 \right)$$

denotes a fBm with Hurst index $\frac{1}{2}(1 - \frac{1}{\alpha})$.

As in the case $\alpha = 2$, we can precise the fractional Brownian motion which appears in the decomposition (5.51) of the solution. We set, for every $t \geq 0, x \in \mathbb{R}$,

$$U_\alpha(t,x) \qquad (5.52)$$

$$= \int_{\mathbb{R}} \int_{\mathbb{R}} (G_\alpha((t-u)_+, x-y) - G_\alpha((-u)_+, x-y)) \tilde{W}(du, dy)$$

with $\tilde{W}$ given by (5.21). Otherwise, we can write

$$U_\alpha(t, x) = \int_{-\infty}^0 \left(G_\alpha(t - u, x - y) - G_\alpha(-u, x - y) \right) \tilde{W}(du, dy)$$

$$+ \int_0^t \int_{\mathbb{R}} G_\alpha(t - u, x - y) W(du, dy)$$

with W the space-time white noise that governs (5.38). We can prove the following result.

Proposition 5.18.

(1) Let U_α be given by (5.52). Then for every $x \in \mathbb{R}$, $(U_\alpha(t, x), t \geq 0)$ is $\frac{1}{2}\left(1 - \frac{1}{\alpha}\right)$-self-similar and it has stationary increments. Consequently the process $(U_\alpha(t, x), t \geq 0)$ has the same law as

$$\left(\sqrt{C_2} B_t^{\frac{1}{2}\left(1 - \frac{1}{\alpha}\right)}, t \geq 0 \right),$$

where $B^{\frac{1}{2}\left(1 - \frac{1}{\alpha}\right)}$ is a fBm with Hurst index $\frac{1}{2}\left(1 - \frac{1}{\alpha}\right)$ and

$$C_2 = \mathbf{E} U_\alpha(1, x)^2 < \infty. \tag{5.53}$$

(2) For $t \geq 0, x \in \mathbb{R}$, let

$$Y_\alpha(t, x) := u(t, x) - U_\alpha(t, x) \tag{5.54}$$

$$= - \int_{-\infty}^0 \int_{\mathbb{R}} \left(G(t - u, x - y) - G(-u, x - y) \right) W(du, dy).$$

Then the mapping $t \to Y_\alpha(t, x) : (0, \infty) \to \mathbb{R}$ is absolutely continuous and of class C^∞ on $(0, \infty)$.

Proof. If suffices to follow the proofs of Proposition 5.8 and of Proposition 5.27. $\qquad\square$

We obtain the following decomposition.

Theorem 5.3. *Let $(u(t, x), t \geq 0, x \in \mathbb{R})$ be the mild solution (5.46). Then, if $x \in \mathbb{R}$ is fixed, we have*

$$(u(t, x), t \geq 0) \equiv^{(d)} \left(\sqrt{C_2} B_t^{\frac{1}{2}\left(1 - \frac{1}{\alpha}\right)} + Y_\alpha(t, x), t \geq 0 \right)$$

with C_2 from (5.53) and $\left(B_t^{\frac{1}{2}\left(1 - \frac{1}{\alpha}\right)}, t \geq 0 \right)$ a fractional Brownian motion with Hurst index $\frac{1}{2}\left(1 - \frac{1}{\alpha}\right)$ and Y is given by (5.54) and it is such that $t \to Y_\alpha(t, x)$ is of class C^∞.

Remark 5.9. We deduce from the above result and (5.51) that the constant $\sqrt{C_2}$ from (5.53) coincides with $c_{2,\alpha} 2^{\frac{1}{2\alpha}}$, with $c_{2,\alpha}$ defined in (5.49).

5.2.3 *Behavior in space*

Let us now discuss the behavior in space of the solution to the fractional heat equation with additive space-time white noise and the connection with fractional Brownian motion.

Let us consider the random field $(S_\alpha(t, x), t > 0, x \in \mathbb{R})$ defined by

$$S_\alpha(t, x) = \int_t^\infty \int_{\mathbb{R}} \left(G_\alpha(s, y) - G_\alpha(s, x - y) \right) W(ds, dy) \qquad (5.55)$$

with G_α the fundamental solution whose Fourier transform is given by (5.39). First, let us observe that the process S_α with $\alpha \in (1, 2)$, has the same properties as the process S given by (5.31) (which correspond to the case $\alpha = 2$).

Proposition 5.19. *Let S_α be given by (5.55). Then $(S_\alpha(t, x), t \geq 0, x \in \mathbb{R})$ is a well-defined centered Gaussian random field and in addition we have:*

(1) For every $t > 0, x, y \in \mathbb{R}$

$$\mathbf{E} \left| S_\alpha(t, x) - S_\alpha(t, y) \right|^2 \leq C |x - y|^2. \qquad (5.56)$$

(2) For $t > 0$ fixed, the mapping $x \to S_\alpha(t, x)$ is absolutely continuous and it has C^∞-sample paths.

Proof. The proof follows the lines of the proof of Proposition 5.11. We note that for every $t > 0, x \in \mathbb{R}$, via (5.10) and (5.11),

$$\mathbf{E} |S_\alpha(t, x)|^2 = \frac{1}{2\pi} \int_t^\infty ds \int_{\mathbb{R}} d\xi e^{-2s|\xi|^\alpha} \left| 1 - e^{-ix\xi} \right|^2$$

$$= \frac{2}{\pi} \int_0^\infty (1 - \cos(x\xi)) \int_t^\infty ds e^{-2s|\xi|^\alpha}.$$

We calculate the integral ds to get

$$\mathbf{E} |S_\alpha(t, x)|^2 = \frac{1}{\pi} \int_0^\infty d\xi \frac{1 - \cos(x\xi)}{|\xi|^\alpha} e^{-2t|\xi|^\alpha} < \infty.$$

Then, for every $x_1, x_2 \in \mathbb{R}$ and with $t > 0$ fixed, by using

$$|1 - \cos(x_1 - x_2)\xi| \leq |x_1 - x_2|^2 |\xi|^2,$$

we found

$$\mathbf{E} \left| S_\alpha(t, x_2) - S(t, x_1) \right|^2 = \frac{1}{\pi} \int_0^\infty d\xi e^{-2t|\xi|^\alpha} \frac{1 - \cos((x_1 - x_2)\xi)}{|\xi|^\alpha}$$

$$\leq \frac{1}{\pi} |x_1 - x_2|^2 \int_0^\infty d\xi e^{-2t|\xi|^\alpha} |\xi|^{2-\alpha}$$

$$= C |x_1 - x_2|^2$$

with $C = \frac{1}{\pi} \int_0^\infty d\xi |\xi|^{2-\alpha} e^{-2t|\xi|^\alpha} < \infty$. Similarly, by proceeding as before in the proof of Proposition 5.11, we see that the mapping $x \to S_\alpha(t, x)$ is absolutely continuous and it has C^∞ sample paths. $\qquad\square$

Let us set, for $t \geq 0, x \in \mathbb{R}$,

$$V_\alpha(t, x) = u(t, x) - S_\alpha(t, x) \qquad (5.57)$$

with u being the mild solution to the fractional heat equation given by (5.46) and S_α defined by (5.55).

Proposition 5.20. *Let V_α be given by (5.57). Then for every fixed $t > 0$, the stochastic process $(V_\alpha(t, x), x \in \mathbb{R})$ coincides in distribution with*

$$\left(\sqrt{C_{0,\alpha}} B^{\frac{\alpha-1}{2}}(x), x \in \mathbb{R} \right)$$

where $(B^{\frac{\alpha-1}{2}}(x), x \in \mathbb{R})$ is a fractional Brownian motion (over the whole real line) with Hurst parameter $\frac{\alpha-1}{2} \in (0, \frac{1}{2})$ and

$$C_{0,\alpha} = \frac{1}{\pi} \int_0^\infty \frac{1 - \cos(\xi)}{|\xi|^\alpha} d\xi < \infty. \qquad (5.58)$$

Proof. We write

$$\mathbf{E}\left(V_\alpha(t, x) - V_\alpha(t, y)\right)^2 = \frac{1}{\pi} \int_0^\infty \int_\mathbb{R} \left(1 - \cos((x-y)\xi)\right) e^{-2s|\xi|^\alpha} d\xi ds$$

$$= \frac{1}{2\pi} \int_\mathbb{R} \frac{1 - \cos((x-y)\xi)}{|\xi|^\alpha} d\xi$$

$$= \frac{1}{\pi} \int_0^\infty \frac{1 - \cos((x-y)\xi)}{|\xi|^\alpha} d\xi$$

and with the change of variable $\tilde{\xi} = (x-y)\xi$,

$$\mathbf{E}\left(V_\alpha(t, x) - V_\alpha(t, y)\right)^2 = \frac{1}{\pi} \int_0^\infty \frac{1 - \cos(|\xi|)}{|\xi|^\alpha} d\xi |x - y|^{\alpha-1}$$

$$= C_{0,\alpha} |x - y|^{\alpha-1}$$

with $C_{0,\alpha}$ given by (5.58). Hence, for any $x, y \in \mathbb{R}$,

$$\mathbf{E}V_\alpha(t, x)V_\alpha(t, y)$$

$$= \frac{1}{2} \left[\mathbf{E}\left(V_\alpha(t, x) - V_\alpha(t, y)\right)^2 - \mathbf{E}V_\alpha(t, x)^2 - \mathbf{E}V_\alpha(t, y)^2 \right]$$

$$= \frac{C_{0,\alpha}}{2} \left(|x|^{\alpha-1} + |y|^{\alpha-1} - |x - y|^{\alpha-1} \right)$$

and therefore $(V_\alpha(t, x), x \in \mathbb{R})$ is (modulo a constant) a fractional Brownian motion with Hurst parameter $\frac{\alpha-1}{2}$. $\qquad\square$

We have the following decomposition of the solution in space.

Theorem 5.4. *Let u be given by (5.46) and assume that $t \geq 0$ is fixed. Then we have*

$$(u(t,x), x \in \mathbb{R}) \equiv^{(d)} \left(\sqrt{C_{0,\alpha}} B^{\frac{\alpha-1}{2}}(x) + S_\alpha(t,x), x \in \mathbb{R} \right)$$

where $(B^{\frac{\alpha-1}{2}}(x), x \in \mathbb{R})$ is a two-sided Brownian motion, S is given by (5.31) and $C_{0,\alpha}$ from (5.58). Moreover, for every $t > 0$, the random fields

$$(u(t,x), c \in \mathbb{R}) \text{ and } (S(t,x), x \in \mathbb{R}) \text{ are independent.}$$

Proof. It follows immediately from Proposition 5.20 and the relation (5.57). □

We obtain the following sharp estimate for the spatial increment on the random field (5.46).

Proposition 5.21. *Let $(u(t,x), t \geq 0, x \in \mathbb{R})$ be defined by (5.46). Then for every $t > 0, x, y \in \mathbb{R}$ with $|x - y|$ small enough, there exist two constants $0 < C_1 < C_2$, such that*

$$C_1 |x - y|^{\alpha-1} \leq \mathbf{E}|u(t,x) - u(t,y)|^2 \leq C_2 |x - y|^{\alpha-1}. \tag{5.59}$$

in particular, for $t > 0$ fixed, the application $x \to u(t,x)$ is Hölder continuous of order δ for every $\delta \in \left(0, \frac{\alpha-1}{2}\right)$.

Proof. From (5.57), and since $(u(t,x), x \in \mathbb{R})$ and $(S_\alpha(t,x), x \in \mathbb{R})$ are independent,

$$\mathbf{E}|V_\alpha(t,x) - V_\alpha(t,y)|^2 = \mathbf{E}|u(t,x) - u(t,y)|^2$$
$$+ \mathbf{E}|S_\alpha(t,x) - S_\alpha(t,y)|^2$$

so

$$\mathbf{E}|u(t,x) - u(t,y)|^2 \leq \mathbf{E}|V_\alpha(t,x) - V_\alpha(t,y)|^2 = C_{0,\alpha}|x - t|^{\alpha-1}.$$

Also, using the estimate (5.56) for the field S_α,

$$\mathbf{E}|u(t,x) - u(t,y)|^2 = \mathbf{E}|V_\alpha(t,x) - V_\alpha(t,y)|^2 - \mathbf{E}|S_\alpha(t,x) - S_\alpha(t,y)|^2$$
$$\geq C_{0,\alpha}|x - y|^{\alpha-1} - C|x - y|^2$$

and for $|x - y|$ small enough, since $\alpha - 1 < 2$,

$$\mathbf{E}|u(t,x) - u(t,y)|^2 \geq C_1 |x - y|^{\alpha-1}.$$

The Hölder property follows from the upper bound in (5.59) and the Kolmogorov continuity theorem. □

Remark 5.10.

(1) The spatial Hölder regularity goes up to the order $\frac{\alpha-1}{2}$ while the time Hölder regularity is at maximum $\frac{\alpha-1}{2\alpha}$. So the random field u is α-times more regular in space than in time. For $\alpha = 2$, compare with Remark 5.6.

(2) The double inequality (5.59) shows that u cannot be δ-Hölder in space with $\delta > \frac{\alpha-1}{2}$, see Remark 5.3. As well, by (5.50), u cannot be δ-Hölder in time with $\delta > \frac{\alpha-1}{2\alpha}$.

As in the case of the space-time white-noise, we can obtain the existence of a bicontinuous version for the random field (5.46).

Proposition 5.22. *Let u be given by (5.46). For every $x, y \in \mathbb{R}$, $s, t \geq 0$, we have for $p \geq 2$,*

$$\mathbf{E} \left|u(t,x) - u(s,y)\right|^p \leq C_p \left[|t - s|^{\frac{p(\alpha-1)}{2\alpha}} + |x - y|^{\frac{p(\alpha-1)}{2}}\right].$$

Consequently the trajectories $(t, x) \to u(t, x)$ are almost surely continuous on $(0, \infty) \times \mathbb{R}$.

Proof. The proof is similar to the proof of Proposition 5.20, via the estimates (5.50) and (5.59). □

Chapter 6

The stochastic heat equation with correlated noise in space

6.1 The standard stochastic heat equation with white-colored noise

In Chapter 5 we have seen that the stochastic heat equation driven by the additive space-time white noise admits a random field solution if and only if the spatial dimension is $d = 1$. This is available for both the standard heat equation (5.3) and for the fractional heat equation (5.38). This hypothesis $d = 1$ is pretty restrictive for many practical applications. A possibility to overcome this restriction is to replace the space-time white noise by a Gaussian noise with correlation in time or in space. In this chapter we will consider a spatial correlation for the noise. There are various way to describe this spatial correlation (some kernels are listed in Section 3.4). Our analysis will focus on the case of the Riesz kernel.

6.1.1 *General properties*

In this chapter, we will analyze the situation when the driving noise of the stochastic heat equation is the white-colored Gaussian noise described in Section 3.4, which behaves as a Wiener process in time and as a fractional Brownian sheet in space, see Remark 3.6. Its spatial covariance is given by the so-called Riesz kernel. More precisely, we will consider the stochastic heat equation

$$\frac{\partial}{\partial t}u(t,x) = \frac{1}{2}\Delta u(t,x) + \dot{W}^\gamma(t,x), \quad t \geq 0, x \in \mathbb{R}^d. \qquad (6.1)$$

with

$$u(0,x) = 0 \text{ for every } x \in \mathbb{R}^d.$$

In (5.38), Δ denotes the standard Laplacian on $\mathbb{R}^d$ given by (5.2) and W^γ is the white-colored noise, i.e. $\left(W^\gamma(t,A), t \geq 0, A \in \mathcal{B}(\mathbb{R}^d)\right)$ is a centered

Gaussian field with covariance

$$\mathbf{E}W^\gamma(t, A)W^\gamma(s, B) = (t \wedge s) \int_A \int_B f(x - y)dxdy$$

where f is the Riesz kernel of order $\gamma \in (0, d)$ given by (3.13). Let us recall that we can associate the Hilbert space $\mathcal{P}$ with the white-colored noise W^γ, see Section 4.5. Then the Wiener integral

$$\int_0^\infty \int_{\mathbb{R}^d} \varphi(s, y)W^\gamma(ds, dy)$$

is well-defined for $\varphi \in \mathcal{P}$. In particular, it is also well-defined for $\varphi \in |\mathcal{P}|$ where $|\mathcal{P}|$ is the subset of $\mathcal{P}$ defined by (4.21). We also recall that $|\mathcal{P}|$ contains only functions while $\mathcal{P}$ may contain distributions.

We know that the Riesz kernel f from (3.13) is the Fourier transform of the measure (see e.g. [Treves (1975)])

$$\mu(d\xi) = \|\xi\|^{-\gamma}d\xi \tag{6.2}$$

in the sense that

$$f(x) = \int_{\mathbb{R}^d} e^{i\langle x, \xi \rangle}\mu(d\xi)$$

or, equivalently,

$$\int_{\mathbb{R}^d} f(x)\varphi(x)dx = \int_{\mathbb{R}^d} \mathcal{F}\varphi(\xi)\mu(d\xi) \text{ for every } \varphi \in \mathcal{S}(\mathbb{R}^d).$$

We have the following Parseval type identity

$$\int_{\mathbb{R}^d} \int_{\mathbb{R}^d} \varphi(x)f(x - y)\psi(y)dxdy = \int_{\mathbb{R}^d} \mathcal{F}\varphi(\xi)\overline{\mathcal{F}\psi(\xi)}\mu(d\xi). \tag{6.3}$$

for any $\varphi, \psi \in \mathcal{S}(\mathbb{R}^d)$ (the Schwarz space on $\mathbb{R}^d$) with μ from (6.2). The above relation can be extended to the closure of $\mathcal{S}(\mathbb{R}^d)$ with respect to the inner product

$$\langle \varphi, \psi \rangle := \int_{\mathbb{R}^d} \int_{\mathbb{R}^d} \varphi(x)f(x - y)\psi(y)dxdy.$$

This implies that for all $\varphi, \psi \in \mathcal{P}$ (the Hilbert space associate to the white-colored noise), we have

$$\langle \varphi, \psi \rangle_{\mathcal{P}} = \int_0^\infty \int_{\mathbb{R}^d} \varphi(s, y)\psi(s, z)f(y - z)dydzds$$

$$= \int_0^\infty ds \left(\int_{\mathbb{R}^d} d\xi \mathcal{F}\varphi(s, \cdot)(\xi)\overline{\mathcal{F}\psi(s, \cdot)}(\xi)\|\xi\|^{-\gamma} \right)$$

with f from (3.13) and μ from (6.2).

As usual, the mild solution to (5.38) is given by

$$u(t, x) = \int_0^t \int_{\mathbb{R}^d} G(t - u, x - z) W^\gamma(du, dz) \tag{6.4}$$

where the above integral $W^\gamma(du, dz)$ is a Wiener integral with respect to the Gaussian noise W^γ introduced in Section 4.5 and G is the fundamental solution (5.5).

Let us first give the necessary and sufficient condition for the existence of the mild solution.

Proposition 6.1. *The mild solution (6.4) to the heat equation (6.1) is well-defined if and only if*

$$d < 2 + \gamma \tag{6.5}$$

Under (6.5), for every $t > 0, x \in \mathbb{R}^d$,

$$u(t, x) \sim N\left(0, C_\gamma t^{1 - \frac{d - \gamma}{2}}\right) \tag{6.6}$$

with

$$C_\gamma = \left(1 - \frac{d - \gamma}{2}\right)^{-1} \int_{\mathbb{R}^d} d\xi e^{-\|\xi\|^2} \|\xi\|^{-\gamma} < \infty. \tag{6.7}$$

Moreover, for every $T > 0$ and every $p \geq 1$,

$$\sup_{t \in [0,T], x \in \mathbb{R}} \mathbf{E}|u(t, x)|^p < \infty. \tag{6.8}$$

Proof. Using the isometry (4.22) and the identity (6.3), we have for $t \geq 0, x \in \mathbb{R}^d$, with the constant $g_{\gamma, d}$ from (3.13),

$$\mathbf{E}u(t, x)^2 = g_{\gamma, d} \int_0^t du \int_{\mathbb{R}^d} \int_{\mathbb{R}^d} dy dz G(t - u, x - z) G(t - u, x - y) \|y - z\|^{-(d - \gamma)}$$

$$= \int_0^t du \int_{\mathbb{R}^d} d\xi e^{-u\|\xi\|^2} \|\xi\|^{-\gamma}$$

$$= C_{1,\gamma} \int_0^t du u^{-\frac{d - \gamma}{2}} \tag{6.9}$$

with $C_{1,\gamma} = \int_{\mathbb{R}^d} d\xi e^{-\|\xi\|^2} \|\xi\|^{-\gamma} < \infty$. The integral du is finite if and only if

$$1 - \frac{d - \gamma}{2} > 0$$

which implies (6.5). Assume (6.5) holds true. Then we obtain from (6.9), with C_γ from (6.7),

$$\mathbf{E}u(t, x)^2 = C_\gamma t^{1 - \frac{d - \gamma}{2}}$$

and thus we have (6.6). The last bound in the statement is also trivial from the above computations and (6.6). $\qquad \square$

Remark 6.1.

(1) Condition (6.5) allows in principle to consider the fractional stochastic heat equation with spatially correlated noise in any spatial dimension.
(2) The case $\gamma = 0$ correspond to the space-time white noise. In this case, (6.5) reads $d < 2$ so $d = 1$.

Concerning the covariance (with respect to its two-variables t and x) of the mild solution (6.4), we have for every $0 \leq s \leq t$ and for every $x, y \in \mathbb{R}^d$,

$$\mathbf{E}u(t,x)u(s,y)$$

$$= g_{\gamma,d} \int_0^{t\wedge s} \int_{\mathbb{R}^d} \int_{\mathbb{R}^d} G(t-u, x-z) G(s-u, y-z) f(y-z) \, dz \, dy \, du$$

$$= \int_0^s du \int_{\mathbb{R}^d} d\xi e^{-\frac{1}{2}(t-u)\|\xi\|^2} e^{-\frac{1}{2}(s-u)\|\xi\|^2} e^{-i\langle \xi, x-y \rangle} \|\xi\|^{-\gamma}$$

where we used (5.10) and (5.39). Since

$$\int_0^s du \, e^{-\frac{1}{2}(t+s-2u)\|\xi\|^2} = \frac{1}{\|\xi\|^2} \left(e^{-\frac{1}{2}(t-s)\|\xi\|^2} - e^{-\frac{1}{2}(t+s)\|\xi\|^2} \right)$$

we obtain

$$\mathbf{E}u(t,x)u(s,y) \tag{6.10}$$

$$= \int_{\mathbb{R}^d} d\xi e^{-i\langle \xi, x-y \rangle} \|\xi\|^{-\gamma-2} \left(e^{-\frac{1}{2}(t-s)\|\xi\|^2} - e^{-\frac{1}{2}(t+s)\|\xi\|^2} \right).$$

We can already notice the stationarity of the solution with respect to the spatial variable, but we cannot get a much more close formula as in the case of the space-time white noise. We will explicitly compute the above covariance when one variable (the time or the space variable) is fixed.

Below in this section, we assume (6.5).

6.1.2 *Behavior in time*

We compute the temporal covariance of the solution and we deduce its relation with the bifractional Brownian motion.

Proposition 6.2. *Let $d < 2 + \gamma$ let u be given by (6.4). Assume $x \in \mathbb{R}^d$ is fixed. Then for every $s, t \geq 0$,*

$$\mathbf{E}u(t,x)u(s,y) = C_0' \left((t+s)^{-\frac{d-\gamma}{2}+1} - (t-s)^{-\frac{d-\gamma}{2}+1} \right) \tag{6.11}$$

with

$$C_0' = \frac{1}{-\frac{d-\gamma}{2}+1} \int_{\mathbb{R}^d} d\xi \|\xi\|^{-\gamma} e^{-\frac{1}{2}\|\xi\|^2}.$$

Consequently, the process $(u(t, x), t \geq 0)$ *has the same law as*

$$\left(c_{2,\gamma} B_t^{\frac{1}{2}, 1 - \frac{d-\gamma}{2}} \right)_{t \geq 0}$$

where $\left(B_t^{\frac{1}{2}, 1 - \frac{d-\gamma}{2}}, t \geq 0 \right)$ *is a bifractional Brownian motion with* $H = \frac{1}{2}, K = 1 - \frac{d-\gamma}{2}$ *and*

$$c_{2,\gamma}^2 = 2^{1 - \frac{d-\gamma}{2}} C_0'. \tag{6.12}$$

Proof. For $s, t \geq 0$, $s \leq t$ we write, with $g_{\gamma, d}$ from (3.13),

$$\mathbf{E} u(t, x) u(s, x)$$
$$= g_{\gamma, d} \int_0^{t \wedge s} du \int_{\mathbb{R}^d} \int_{\mathbb{R}^d} G(t - u, x - y) G(s - u, x - y') f(y - y') dy dy'$$
$$= \int_0^s du \int_{\mathbb{R}^d} \mu(d\xi) \mathcal{F} G(t - u, x - \cdot)(\xi) \overline{\mathcal{F} G(s - u, x - \cdot)}(\xi)$$
$$= \int_0^s du \int_{\mathbb{R}^d} d\xi \|\xi\|^{-\gamma} e^{-\frac{1}{2}(t - u)\|\xi\|^2} e^{-\frac{1}{2}(s - u)\|\xi\|^2}. \tag{6.13}$$

From (6.13), by the change of variables $\tilde{\xi} = \sqrt{t + s - 2u} \xi$

$$\mathbf{E} u(t, x) u(s, x)$$
$$= \int_0^s du (t + s - 2u)^{-\frac{d-\gamma}{2}} \int_{\mathbb{R}^d} \mu(d\xi) e^{-\frac{1}{2}\|\xi\|^2}$$
$$= \int_{\mathbb{R}^d} d\xi \|\xi\|^{-\gamma} e^{-\frac{1}{2}\|\xi\|^2} \frac{1}{-\frac{d-\gamma}{2} + 1} \left((t + s)^{-\frac{d-\gamma}{2} + 1} - (t - s)^{-\frac{d-\gamma}{2} + 1} \right)$$
$$= C_0' \left((t + s)^{-\frac{d-\gamma}{2} + 1} - (t - s)^{-\frac{d-\gamma}{2} + 1} \right).$$

To finish the proof, compare the last line above with the covariance of the bifractional Brownian motion (2.20). $\qquad \square$

Remark 6.2.

(1) Condition (6.5) ensures that the index $K = 1 - \frac{d-\gamma}{2}$ of the bifractional Brownian motion is strictly positive. Clearly, $K = 1 - \frac{d-\gamma}{2} < 1$ since $\gamma < d$.

(2) If $d = 1$ and $\gamma = 0$ (that is, we are in the space-time white noise situation), then $H = K = \frac{1}{2}$. Moreover, the constant $c_{2,\gamma}$ in (6.12) becomes

$$c_{2,\gamma}^2 = \sqrt{2} \int_{\mathbb{R}} d\xi e^{-\frac{\|\xi\|^2}{2}} = 2\sqrt{\pi}.$$

(Why we don't retrieve the constant from (5.18) when $\gamma = 0$ and $d = 1$? because the constant $g_{\gamma,d}$ from (3.13) does not reduce to 1 for $\gamma = 0$ and $d = 1$; it is actually not defined for $\gamma = 0$.) As a consequence of the above result, and using the properties of the bifractional Brownian motion discussed in Section 2.2, we notice, by assuming that u is given by (6.4) and (6.5) holds true:

- For every $x \in \mathbb{R}^d$ fixed, the Gaussian process $(u(t,x), t \geq 0)$ is a self-similar process of order $\frac{1}{2}\left(1 - \frac{d-\gamma}{2}\right)$.
- For $x \in \mathbb{R}^d$ fixed, the sample path $t \to u(t,x)$ is (modulo a modification) Hölder continuous of order δ for every $\delta \in \left(0, \frac{1}{2}\left(1 - \frac{d-\gamma}{2}\right)\right)$.
- Moreover, there exist two constants $0 < C_1 < C_2$ (they may depend on d, γ) such that

$$C_1 |t - s|^{1 - \frac{d-\gamma}{2}} \leq \mathbf{E}|u(t,x) - u(s,x)|^2 \leq C_2 |t - s|^{1 - \frac{d-\gamma}{2}}$$

for every $s, t \geq 0, x \in \mathbb{R}^d$.

An "explicite" decomposition in law of the solution can be also given as in Theorems 5.1 and 5.3. Define for $t \geq 0, x \in \mathbb{R}^d$,

$$
\begin{aligned}
U_\gamma(t,x) \\
&= \int_{\mathbb{R}} \int_{\mathbb{R}^d} \left(G((t-s)_+, x-y) - G((-s)_+, x-y) \right) \tilde{W}^\gamma(ds, dy) \\
&= \int_0^t \int_{\mathbb{R}^d} G(t-s, x-y) W^\gamma(ds, dy) \\
&\quad + \int_{-\infty}^0 \int_{\mathbb{R}^d} \left(G(t-s, x-y) - G(-s, x-y) \right) \tilde{W}^\gamma(ds, dy). \quad (6.14)
\end{aligned}
$$

Above $\left(\tilde{W}^\gamma(t, A), t \in \mathbb{R}, A \in \mathcal{B}_b(\mathbb{R}^d) \right)$ is defined, for $t \in \mathbb{R}, A \in \mathcal{B}_b(\mathbb{R}^d)$, by

$$\tilde{W}^\gamma(t, A) = \begin{cases} W^\gamma(t, A) & \text{if } t \geq 0. \\ W^\gamma(-t, A) & \text{if } t < 0. \end{cases}$$

Its covariance satisfies, for $s, t \in \mathbb{R}, A, B \in \mathcal{B}_b(\mathbb{R}^d)$,

$$\mathbf{E}\tilde{W}^\gamma(t, A)\tilde{W}^\gamma(s, B) = \frac{1}{2}(|t| + |s| - |t - s|) \int_A \int_B f(y - z) dy dz. \quad (6.15)$$

We can naturally define a Wiener integral with respect to $\tilde{W}^\gamma$ as in Section 4.5.

Proposition 6.3. *Assume (6.5).*

(1) *Let U_γ be given by (6.14) with $x \in \mathbb{R}^d$ fixed. Then the stochastic process $(U_\gamma(t,x), t \geq 0)$ is $\frac{1}{2}\left(1 - \frac{d-\gamma}{2}\right)$-self similar and it has stationary increments. Consequently, $(U(t,x), t \geq 0)$ has the same finite dimensional distributions as*

$$\left(\sqrt{C_3} B^{\frac{1}{2}\left(1 - \frac{d-\gamma}{2}\right)}, t \geq 0\right)$$

with $\left(B^{\frac{1}{2}\left(1 - \frac{d-\gamma}{2}\right)}, t \geq 0\right)$ a fBm with Hurst index $\frac{1}{2}\left(1 - \frac{d-\gamma}{2}\right)$ and $C_3 = \mathbf{E}U(1,x)^2$.

(2) *For $t \geq 0, x \in \mathbb{R}^d$, let*

$$Y_\gamma(t,x) = u(t,x) - U_\gamma(t,x).$$

Then the sample path $t \to Y_\gamma(t,x)$ is absolutely continuous of class C^∞ on $(0, \infty)$.

Proof. Although, the proof of self-similarity can be done by following the proof of Theorem 5.8, we will employ a slightly different argument which does not use the Gaussian character of U_γ. We can assume $x = 0$, because the processes $(U_\gamma(t,x), t \geq 0)$ and $(U_\gamma(t,0), t \geq 0)$ have the same finite dimensional distributions. For $c > 0$, we write

$$
\begin{aligned}
&U_\gamma(ct, 0)\\
&= \int_\mathbb{R} \int_{\mathbb{R}^d} \left(G((ct-s)_+, -y) - G((-s)_+, x-y)\right) \tilde{W}^\gamma(ds, dy)\\
&= \int_\mathbb{R} \int_{\mathbb{R}^d} \left(G(c(t-s)_+, -y) - G((-cs)_+, -y)\right) \tilde{W}^\gamma\left(d\left(\frac{s}{c}\right), dy\right)\\
&\equiv^{(d)} c^{-\frac{1}{2}} \int_\mathbb{R} \int_{\mathbb{R}^d} \left(G(c(t-s)_+, -y) - G((-cs)_+, -y)\right) \tilde{W}^\gamma(ds, dy).
\end{aligned}
$$

Now we use the scaling property of the Green kernel (5.43) to write

$$
\begin{aligned}
&U_\gamma(ct, 0)\\
&\equiv^{(d)} c^{-\frac{1}{2}} c^{-\frac{d}{2}} \int_\mathbb{R} \int_{\mathbb{R}^d} \tilde{W}^\gamma(ds, dy)\\
&\quad \times \left(G((t-s)_+, -c^{-\frac{1}{2}}y) - G((-s)_+, -c^{-\frac{1}{2}}y)\right)\\
&= c^{-\frac{1}{2}} c^{-\frac{d}{2}} \int_\mathbb{R} \int_{\mathbb{R}^d} \left(G((t-s)_+, -y) - G((-s)_+, -y)\right) \tilde{W}^\gamma\left(ds, d(c^{\frac{1}{2}}y)\right)\\
&\equiv^{(d)} c^{-\frac{1}{2}} c^{-\frac{d}{2}} c^{\frac{d}{2} \frac{1+\gamma}{2}} \int_\mathbb{R} \int_{\mathbb{R}^d} \left(G((t-s)_+, -y) - G((-s)_+, -y)\right) \tilde{W}^\gamma(ds, dy)\\
&= c^{\frac{1}{2}\left(1 - \frac{d-\gamma}{2}\right)} U_\gamma(t, 0).
\end{aligned}
$$

where we used the spatial scaling of the white-colored noise W^γ. The stationarity of the increments of U in time and point 2. of the statement can be done on in Theorem 5.8.   □

Theorem 6.1. *Let u be the mild solution and fix $x \in \mathbb{R}^d$. Then*

$$\left(u(t,x), t \geq 0\right) \equiv^{(d)} \left(\sqrt{C_3} B_t^{\frac{1}{2}\left(1-\frac{d-\gamma}{2}\right)} + Y_\gamma(t,x), t \geq 0\right)$$

with $\left(B_t^{\frac{1}{2}\left(1-\frac{d-\gamma}{2}\right)}, t \geq 0\right)$ a fBm with index $\frac{1}{2}\left(1 - \frac{d-\gamma}{2}\right)$ and $t \to Y_\gamma(t,x)$ of class C^∞ on $(0, \infty)$.

Remark 6.3. We deduce from Proposition 6.3 and Proposition 6.2 that the constant $\sqrt{C_3}$ coincides with $c_{2,\gamma} 2^{\frac{1}{2} - \frac{d-\gamma}{4}}$ with $c_{2,\gamma}$ in (6.12).

6.1.3   *Behavior in space*

A spatial decomposition similar to that obtained for $\gamma = 0$ can be given. The main difference with respect to the space-time white noise situation is that now the spatial dimension is not equal to 1 anymore and we will see the appearance of the isotropic fractional Brownian sheet.

Let us put for every $t > 0, x \in \mathbb{R}^d$,

$$S_\gamma(t,x) = \int_t^\infty \int_{\mathbb{R}^d} \left(G(s,z) - G(s, x-z)\right) W^\gamma(ds, dz). \qquad (6.16)$$

In the next result, we give some properties of the random field S. As for $\gamma = 0$, the process S_γ has nice regularity properties with respect to its spatial variable.

Proposition 6.4. *Let S_γ be defined by (6.16). Then*

- *The random field $(S_\gamma(t,x), t > 0, x \in \mathbb{R}^d)$ is well-defined.*
- *For every $t > 0, x_1, x_2 \in \mathbb{R}^d$,*

$$\mathbf{E} \left|S_\gamma(t, x_1 - S_\gamma(t, x_2)\right|^2 \leq C\|x_1 - x_2\|^2.$$

- *For $t \geq 0$ fixed, the mapping $x \to S_\gamma(t,x)$ from $\mathbb{R}^d$ onto $\mathbb{R}$ is of class C^∞.*

Proof. For $t \geq 0, x \in \mathbb{R}^d$, using the Wiener isometry (4.22) and (6.3)

$$\mathbf{E}S_\gamma(t,x)^2$$

$$= \int_t^\infty ds \int_{\mathbb{R}^d} \int_{\mathbb{R}^d} dy dz \, (G(s,y) - G(s,x-y)) \, (G(s,z) - G(s,x-z)) \, f(y-z)$$

$$= \int_t^\infty ds \int_{\mathbb{R}^d} d\xi \|\xi\|^{-\gamma} \, (\mathcal{F}G(s,\cdot)(\xi) - \mathcal{F}G(s,x-\cdot)(\xi))$$

$$\times \overline{(\mathcal{F}G(s,\cdot)(\xi) - \mathcal{F}G(s,x-\cdot)(\xi))}$$

$$= \int_t^\infty du \int_{\mathbb{R}^d} d\xi \|\xi\|^{-\gamma} e^{-u\|\xi\|^2} \left| 1 - e^{-i\langle\xi,x\rangle} \right|^2$$

We compute first the integral ds to find

$$\mathbf{E}S(t,x)^2 = \int_{\mathbb{R}^d} d\xi \|\xi\|^{-\gamma-2}(1 - \cos(\langle\xi,x\rangle))e^{-t\|\xi\|^2}$$

$$= \int_{\|\xi\|\leq 1} d\xi \|\xi\|^{-\gamma-2}(1 - \cos(\langle\xi,x\rangle))e^{-t\|\xi\|^2}$$

$$+ \int_{\|\xi\|\geq 1} d\xi \|\xi\|^{-\gamma-2}(1 - \cos(\langle\xi,x\rangle))e^{-t\|\xi\|^2}.$$

By majorizing the exponential function by 1 and using $|1 - \cos(\langle\xi,x\rangle)| \leq C\|\xi\|^2\|x\|^2$, the integral over the region $\|\xi\| \leq 1$ is bounded by

$$C\|x\|^2 \int_{\|\xi\|\leq 1} d\xi \|\xi\|^{-\gamma}$$

which is finite because $\gamma < d$. For the integral over the region $\|\xi\| > 1$ we use $|1 - \cos(\langle\xi,x\rangle)| \leq 2$ and we see that this integral is also finite due to the presence of the exponential function. Thus $\mathbf{E}S_\gamma(t,x)^2 < \infty$ so S_γ is a well-defined random field.

For point 2., we write for $x_1, x_2 \in \mathbb{R}^d$, again via Parseval (6.3), and as in the proof of Proposition 5.11,

$$\mathbf{E}\left|S_\gamma(t,x_1) - S_\gamma(t,x_2)\right|^2$$

$$= \int_t^\infty du \int_{\mathbb{R}^d} d\xi \|\xi\|^{-\gamma} e^{-u\|\xi\|^2} \left| 1 - e^{-i\langle\xi,(x_1-x_2)\rangle} \right|^2$$

$$= 2 \int_t^\infty du \int_{\mathbb{R}^d} d\xi \|\xi\|^{-\gamma} e^{-u|\xi|^2} (1 - \cos(\langle\xi, x_1 - x_2\rangle))$$

$$= 2 \int_{\mathbb{R}^d} d\xi \|\xi\|^{-\gamma-2} e^{-t\|\xi\|^2} (1 - \cos(\langle\xi, x_1 - x_2\rangle))$$

where we used Fubini and we calculated the integral du. Now, using $|1 - \cos(\langle\xi, x_1 - x_2\rangle)| \leq \|x_1 - x_2\|^2\|\xi\|^2$, we can write

$$\mathbf{E}\left|S_\gamma(t,x_1) - S_\gamma(t,x_2)\right|^2 \leq 2(\|x_1 - x_2\|^2 \int_{\mathbb{R}^d} d\xi \|\xi\|^{-\gamma} e^{-u\|\xi\|^2}$$

and we get the conclusion since the integral $\int_{\mathbb{R}^d} d\xi \|\xi\|^{-\gamma} e^{-u\|\xi\|^2}$ is finite for $0 < \gamma < d$.

Concerning 3., let us denote, for $n \geq 1$ and for the positive integers $\alpha_1, \ldots, \alpha_n$ with $\alpha_1 + \ldots + \alpha_n = n$,

$$Y(t,x) = \int_t^\infty \int_{\mathbb{R}^d} \frac{\partial^n}{\partial x_1^{\alpha_1} \ldots \partial x_n^{\alpha_n}} (G(s,y) - G(s, x-y)) W^\gamma(ds, dy).$$

This is the formal derivative of S_γ with respect to $\partial x_1^{\alpha_1} \ldots \partial x_n^{\alpha_n}$. We have

$$\mathbf{E}|Y(t,x)|^2 = \int_t^\infty ds \int_{\mathbb{R}^d} dy \left| \frac{\partial^n}{\partial x_1^{\alpha_1} \ldots \partial x_n^{\alpha_n}} G(s, x-y) \right|^2$$

$$= \int_t^\infty ds \int_{\mathbb{R}^d} d\xi \|\xi\|^{-\gamma} \left| \mathcal{F} \frac{\partial^n}{\partial x_1^{\alpha_1} \ldots \partial x_n^{\alpha_n}} G(s, x - \cdot)(\xi) \right|^2$$

Since

$$\mathcal{F} \frac{\partial^n}{\partial x_1^{\alpha_1} \ldots \partial x_n^{\alpha_n}} G(s, x - \cdot)(\xi) = (i\xi)^n e^{-\frac{s\|\xi\|^2}{2}}$$

we get

$$\left| \mathcal{F} \frac{\partial^n}{\partial x_1^{\alpha_1} \ldots \partial x_n^{\alpha_n}} G(s, x - \cdot)(\xi) \right|^2 = \|\xi\|^{2n} e^{-s\|\xi\|^2}$$

and thus

$$\mathbf{E}|Y(t,x)|^2 = \int_t^\infty ds \int_{\mathbb{R}^d} d\xi \|\xi\|^{2n-\gamma} e^{-s\|\xi\|^2}$$

$$= \int_{\mathbb{R}^d} d\xi \|\xi\|^{2n-2-\gamma} e^{-s\|\xi\|^2} < \infty.$$

This proves the nth times differentiability of $S_\gamma(t,x)$ with respect to $x \in \mathbb{R}^d$ for an arbitrary integer $n \geq 1$. $\square$

Define, for $t \geq 0, x \in \mathbb{R}$,

$$V_\gamma(t,x) = u(t,x) - S_\gamma(t,x). \tag{6.17}$$

Proposition 6.5. *Assume (6.5) and let V_γ be given by (6.17). Let $t > 0$ be fixed. The random field $(V_\gamma(t,x), x \in \mathbb{R}^d)$ coincides in distribution with*

$$\left(\sqrt{C_{0,\gamma}} B^{\frac{2+\gamma-d}{2}}(x), x \in \mathbb{R}^d \right)$$

where $\left(B^{\frac{2+\gamma-d}{2}}(x), x \in \mathbb{R}^d \right)$ is an isotropic fractional Brownian sheet with Hurst index $\frac{2+\gamma-d}{2}$ and

$$C_{0,\gamma} = 2 \int_{\mathbb{R}^d} \|\xi\|^{-\gamma-2} (1 - \cos(\langle \xi, e_1 \rangle)). \tag{6.18}$$

Proof. We have for $x, y \in \mathbb{R}^d$,

$$V_\gamma(t, x) - V_\gamma(t, y)$$

$$= -\int_t^\infty \int_{\mathbb{R}^d} \left(G(u, x - z) - G(u, y - z)\right) W^\gamma(du, dz)$$

$$+ \int_0^t \int_{\mathbb{R}^d} \left(G(u, x - z) - G(u, y - z)\right) W^\gamma(du, dz).$$

Since W^γ behaves as a Brownian motion in time, we can notice that the random variables

$$\int_t^\infty \int_{\mathbb{R}^d} \left(G(u, x - z) - G(u, y - z)\right) W^\gamma(du, dz)$$

and

$$\int_0^t \int_{\mathbb{R}^d} \left(G(u, x - z) - G(u, y - z)\right) W^\gamma(du, dz)$$

are independent. Thus

$$\mathbf{E}|V_\gamma(t, x) - V_\gamma(t, y)|^2$$

$$= \mathbf{E}\left|\int_t^\infty \int_{\mathbb{R}^d} \left(G(u, x - z) - G(u, y - z)\right) W^\gamma(du, dz)\right|^2$$

$$+ \mathbf{E}\left|\int_0^t \int_{\mathbb{R}^d} \left(G(u, x - z) - G(u, y - z)\right) W^\gamma(du, dz)\right|^2$$

$$= \int_0^\infty \int_{\mathbb{R}^d} f(z - z') dz dz' du$$

$$\times \left(G(u, x - z) - G(u, y - z)\right)\left(G(u, x - z') - G(u, y - z')\right)$$

$$= \int_0^\infty \int_{\mathbb{R}^d} d\xi \|\xi\|^{-\gamma} e^{-u\|\xi\|^2} \left|1 - e^{-i\langle \xi, x - y \rangle}\right|^2$$

$$= 2 \int_0^\infty \int_{\mathbb{R}^d} d\xi \|\xi\|^{-\gamma} e^{-u\|\xi\|^2} \left(1 - \cos(\xi, x - y)\right)$$

Now, since

$$\int_0^\infty du\, e^{-u\|\xi\|^2} = \|\xi\|^{-2}$$

we obtain

$$\mathbf{E}|V_\gamma(t, x) - V_\gamma(t, y)|^2 = 2 \int_{\mathbb{R}^d} \|\xi\|^{-\gamma-2} \left(1 - \cos(\xi, x - y)\right)$$

$$= 2\|x - y\|^{2+\gamma-d} \int_{\mathbb{R}^d} \|\xi\|^{-\gamma-2} \left(1 - \cos(\langle \xi, e_1 \rangle)\right)$$

where $e_1 = (1, 0, \ldots, 0) \in \mathbb{R}^d$. Thus, for every $x, y \in \mathbb{R}^d$,

$$\mathbf{E}V_\gamma(t, x)V_\gamma(t, y) = C_{0,\gamma} \frac{1}{2} \left(\|x\|^{2H} + \|y\|^{2H} - \|x - y\|^{2H}\right)$$

with $C_{0,\gamma}$ defined by (6.18) and $H = \frac{2+\gamma-d}{2}$. This gives the conclusion. $\square$

6.2   The fractional heat equation with white-colored noise

6.2.1   General properties

We will add a new parameter to the heat equation (5.38), by considering a Gaussian noise which behaves as a fractional Brownian motion in space, i.e. its spatial covariance is given by the Riesz kernel. More precisely, we will consider the stochastic heat equation

$$\frac{\partial}{\partial t}u(t,x) = -(-\Delta)^{\frac{\alpha}{2}}u(t,x) + \dot{W}^{\gamma}(t,x), \qquad t \geq 0, x \in \mathbb{R}^d. \qquad (6.19)$$

with $u(0,x) = 0$ for every $x \in \mathbb{R}^d$. In (5.38), $-(-\Delta)^{\frac{\alpha}{2}}$ denotes the fractional Laplacian with exponent $\frac{\alpha}{2}$, $\alpha \in (1,2]$ and W^{γ} is the so-called white-colored noise, i.e. $\big(W^{\gamma}(t,A), t \geq 0, A \in \mathcal{B}(\mathbb{R}^d)\big)$ is a centered Gaussian field with covariance

$$\mathbf{E}W^{\gamma}(t,A)W^{\gamma}(s,B) = (t \wedge s)\int_A \int_B f(x-y)dxdy$$

where f is the Riesz kernel of order γ given by (3.13).

As usual, the mild solution to (6.19) is expressed as a Wiener integral with respect to the white-colored noise W^{γ}, i.e.

$$u(t,x) = \int_0^t \int_{\mathbb{R}^d} G_\alpha(t-u, x-y)W^{\gamma}(du, dy) \qquad (6.20)$$

where G_α is the fundamental solution of the fractional heat equation satisfying (5.39) and the above integral is a Wiener integral with respect to W^{γ} as defined in Section 4.5.

Many of the arguments from the previous section can be repeated in the case of the fractional heat equation, i.e. by replacing the standard Laplacian in (6.1) by the fractional Laplacian. We will follow the same steps as in the previous paragraphs: first we give a necessary and sufficient condition for the existence of the mild solution, then we discuss its relation with the fractional and bifractional Brownian motion and we deuce interesting properties.

Let us first give the necessary and sufficient condition for the existence of the mild solution.

Proposition 6.6. *The mild solution (6.4) to the heat equation (6.1) is well-defined if and only if*

$$d < \alpha + \gamma. \qquad (6.21)$$

Under (6.21), for every $t > 0, x \in \mathbb{R}^d$,

$$u(t,x) \sim N\left(0, C_{1,\alpha,\gamma} t^{1-\frac{d-\gamma}{2}}\right)$$

with

$$C_{1,\alpha,\gamma} = \int_{\mathbb{R}^d} d\xi e^{-\|\xi\|^\alpha} \|\xi\|^{-\gamma} \frac{1}{1 - \frac{d-\gamma}{\alpha}} < \infty. \tag{6.22}$$

Moreover, for every $T > 0$ and every $p \geq 1$,

$$\sup_{t \in [0,T], x \in \mathbb{R}} \mathbf{E}|u(t,x)|^p < \infty. \tag{6.23}$$

Proof. Using the Wiener isometry (4.22) and the identity (6.3), we have for $t \geq 0, x \in \mathbb{R}^d$,

$$\begin{aligned}
\mathbf{E}u(t,x)^2 &= g_{\gamma,d} \int_0^t du \int_{\mathbb{R}^d} \int_{\mathbb{R}^d} dydz |y - z|^{-(d-\gamma)} \\
&\quad \times G_\alpha(t - u, x - z) G_\alpha(t - u, x - y) \\
&= \int_0^t du \int_{\mathbb{R}^d} d\xi e^{-u\|\xi\|^\alpha} \|\xi\|^{-\gamma} \\
&= \int_0^t du\, u^{-\frac{d-\gamma}{\alpha}} \int_{\mathbb{R}^d} d\xi e^{-\|\xi\|^\alpha} \|\xi\|^{-\gamma}.
\end{aligned}$$

The integral du is finite if and only if

$$1 - \frac{d - \gamma}{\alpha} > 0 \text{ which is equivalent to } d < \alpha + \gamma$$

and this implies (6.21). If (6.21) holds true, then

$$\mathbf{E}u(t,x)^2 = C_{1,\alpha,\gamma} t^{1-\frac{d-\gamma}{2}}$$

with $C_{1,\alpha,\gamma}$ from (6.22), so $u(t,x) \sim N\left(0, C_{1,\alpha,\gamma} t^{1-\frac{d-\gamma}{2}}\right)$. We clearly have (6.23) for $p = 2$ and then it is obtained for any $p \geq 1$ because of the Gaussian distribution of $u(t,x)$. $\square$

6.2.2 Behavior in time

In the next result we deduce the law of the Gaussian process $u(t,x), t \geq 0$ with $x \in \mathbb{R}^d$ fixed.

Proposition 6.7. *Assume (6.21). For every $s, t \geq 0$, and for every $x \in \mathbb{R}^d$, we have*

$$\mathbf{E}u(t,x)u(s,x) = c_{1,\alpha,\gamma}\left[(t + s)^{1-\frac{d-\gamma}{\alpha}} - |t - s|^{1-\frac{d-\gamma}{\alpha}}\right] \tag{6.24}$$

where

$$c_{1,\alpha,\gamma} = \int_{\mathbb{R}^d} d\xi \|\xi\|^{-\gamma} e^{-\|\xi\|^\alpha} \frac{1}{2(1 - \frac{d-\gamma}{\alpha})}. \tag{6.25}$$

Consequently, the process $(u(t,x), t \geq 0)$ *has the same law as*

$$\left(c_{2,\alpha,\gamma} B_t^{\frac{1}{2}, 1 - \frac{d-\gamma}{\alpha}}, t \geq 0\right)$$

where $\left(B^{\frac{1}{2}, 1 - \frac{d-\gamma}{\alpha}}, t \geq 0\right)$ *is a bi-fBm with* $H = \frac{1}{2}$ *and* $K = 1 - \frac{d-\gamma}{\alpha}$ *and*

$$c_{2,\alpha,\gamma}^2 = c_{1,\alpha,\gamma} 2^{1 - \frac{d-\gamma}{\alpha}}. \tag{6.26}$$

Proof. As in the proof of Proposition 5.16 we have for $0 \leq s \leq t$ and for $x \in \mathbb{R}^d$,

$$\begin{aligned}
\mathbf{E}u(t,x)u(s,x) &= \int_0^{t \wedge s} du \int_{\mathbb{R}^d} d\xi \|\xi\|^{-\gamma} e^{-(t-u)\|\xi\|^\alpha} e^{-(s-u)\|\xi\|^\alpha} \\
&= \int_{\mathbb{R}^d} d\xi \|\xi\|^{-\gamma} e^{-\|\xi\|^\alpha} \int_0^s du (t + s - 2u)^{-\frac{d-\gamma}{\alpha}} \\
&= c_{1,\alpha,\gamma} \left((t+s)^{1 - \frac{d-\gamma}{\alpha}} - |t - s|^{1 - \frac{d-\gamma}{\alpha}}\right).
\end{aligned}$$

with $c_{1,\alpha,\gamma}$ given by (6.25). $\qquad \square$

Let us make some comments.

Remark 6.4.

- The case $\gamma = 0$ in (6.24) means that the noise of the heat equation has no "spatial color", i.e. it is the space-time white noise. So in this case the spatial dimension d is forced to be equal to 1 and we retrouve the formula (5.48).
- The case $\alpha = 2$ in (6.24) corresponds to the situation of the standard Laplacian. In this case, (6.24) reduces, modulo a constant, to (6.11).

We deduce the following properties.

Proposition 6.8. *Let u be given by (6.20) and assume (6.21) holds. Let $x \in \mathbb{R}^d$ be fixed. Then*

(1) *The process* $(u(t,x), t \geq 0)$ *is self-similar with index* $\frac{1}{2}\left(1 - \frac{d-\gamma}{\alpha}\right)$.

(2) *The sample paths of the process* $(u(t,x), t \geq 0)$ *are Hölder continuous of order δ for every* $\delta \in \left(0, \frac{1}{2}\left(1 - \frac{d-\gamma}{\alpha}\right)\right)$.

(3) There exist two constant s $0 < C_1 < C_2$ such that

$$C_1|t - s|^{1 - \frac{d-\gamma}{\alpha}} \le \mathbf{E}|u(t, x) - u(s, x)|^2 \le C_2|t - s|^{1 - \frac{d-\gamma}{\alpha}}$$

for every $s, t \ge 0$.

(4) There exists a Gaussian process $(Y_t, t \ge 0)$ with absolutely continuous and C^∞ sample paths, independent of u, such that

$$\left(c_{2,\alpha,\gamma}^{-1} u(t, x) + Y_t, t \ge 0\right) =^{(d)} \left(\sqrt{2^{1-K}} B^{\frac{1}{2}\left(1 - \frac{d-\gamma}{\alpha}\right)}, t \ge 0\right)$$

where $B^{\frac{1}{2}\left(1 - \frac{d-\gamma}{\alpha}\right)}$ is a fBm with Hurst index $\frac{1}{2}\left(1 - \frac{d-\gamma}{\alpha}\right)$ and $K = \left(1 - \frac{d-\gamma}{\alpha}\right)$.

Proof. The proof is obtained by using the properties of the bifractional Brownian motion with parameters $H = \frac{1}{2}$ and $K = 1 - \frac{d-\gamma}{\alpha}$. □

Define

$$U_{\alpha,\delta}(t, x) = \int_{\mathbb{R}} \int_{\mathbb{R}^d} \left(G((t - s)_+, x - y) - G((-s)_+, x - y)\right) \tilde{W}^\gamma(ds, dy)$$
$$(6.27)$$

with $\tilde{W}^\gamma$ from (6.15). Then, by following the proofs of Propositions 5.8 and 5.9, we have the following:

Proposition 6.9.

(1) Let $U_{\alpha,\delta}$ be given by (6.27) with $x \in \mathbb{R}^d$ fixed. Then the stochastic process $(U_{\alpha,\delta}(t, x), t \ge 0)$ is $\frac{1}{2}\left(1 - \frac{d-\gamma}{\alpha}\right)$-self similar and it has stationary increments. Consequently, $(U_{\alpha,\delta}(t, x), t \ge 0)$ has the same finite dimensional distributions as

$$\left(\sqrt{C_4} B^{\frac{1}{2}\left(1 - \frac{d-\gamma}{\alpha}\right)}, t \ge 0\right)$$

with $\left(B^{\frac{1}{2}\left(1 - \frac{d-\gamma}{\alpha}\right)}, t \ge 0\right)$ a fBm with Hurst index $\frac{1}{2}\left(1 - \frac{d-\gamma}{\alpha}\right)$ and $C_4 = \mathbf{E}U_{\alpha,\delta}(1, x)^2$.

(2) For $t \ge 0, x \in \mathbb{R}^d$, let

$$Y_{\alpha,\delta}(t, x) = u(t, x) - U_{\alpha,\delta}(t, x).$$

Then $t \to Y_{\alpha,\delta}(t, x)$ is of class C^∞ on $(0, \infty)$.

6.2.3 Behavior in space

Now we analyze the process $(u(t, x), x \in \mathbb{R}^d)$ with $t \geq 0$ fixed. We will see that it is related to the anisotropic fractional Brownian sheet discussed in Section 3.3.

The spatial covariance of the process u can be computed from (6.10) by taking $t = s$,

$$\mathbf{E}u(t, x)u(s, y) = \int_{\mathbb{R}^d} d\xi e^{-i\langle \xi, x-y \rangle} \|\xi\|^{-\gamma-\alpha} \left(1 - e^{-t\|\xi\|^\alpha} \right). \quad (6.28)$$

Proposition 6.10. *Let $t > 0$ be fixed. The process $(u(t, x), x \in \mathbb{R}^d)$ is stationary.*

Proof. We deal with a centered Gaussian process and for every $x, y \in \mathbb{R}^d$ and $h \in (0, \infty)n^d$, by (6.28)

$$\mathbf{E}(u(t, x + h)u(t, y + h)) = \mathbf{E}u(t, x)u(t, y).$$

$\square$

We give a spatial decomposition of the random field u as we did in the previous paragraphs. Define, for $x \in \mathbb{R}^d$,

$$S_{\alpha,\delta}(x) = \int_t^\infty \int_{\mathbb{R}^d} [G_\alpha(s, z) - G_\alpha(u, x - z)] \, W^\gamma(ds, dz). \quad (6.29)$$

Then we have the following.

Proposition 6.11. *Let $(S_{\alpha,\delta}(t, x), t \geq 0, x \in \mathbb{R}^d)$ be defined by (6.29). Then*

(1) The random field $S_{\alpha,\delta}$ is well-defined, i.e. for every $t \geq 0, x \in \mathbb{R}^d$

$$\mathbf{E}|S_\alpha(t, x)|^2 < \infty.$$

(2) The paths $x \to S_{\alpha,\delta}(t, x)$ are absolute continuous and of class C^∞.
(3) For every $x_1, x_2 \in \mathbb{R}$,

$$\mathbf{E}\left|S_{\alpha,\delta}(t, x_1) - S_{\alpha,\delta}(t, x_2)\right|^2 \leq C\|x_1 - x_2\|^2. \quad (6.30)$$

Proof. For 1., we have

$$\mathbf{E}|S_{\alpha,\delta}(x)|^2$$

$$= \int_t^\infty ds \int_{\mathbb{R}^d} \int_{\mathbb{R}^d} dz dz'$$
$$\times \left(G_\alpha(s,z) - G_\alpha(s,x-z)\right) \left(G_\alpha(s,z') - G_\alpha(s,x-z')\right) f(z-z')$$

$$= \int_t^\infty du \int_{\mathbb{R}^d} d\xi \|\xi\|^{-\gamma}$$
$$\times \left(\mathcal{F}G_\alpha(s,\cdot)(\xi) - \mathcal{F}G_\alpha(s,x-\cdot)(\xi)\right) \overline{\left(\mathcal{F}G_\alpha(s,\cdot)(\xi) - \mathcal{F}G_\alpha(s,x-\cdot)(\xi)\right)}$$

$$= \left(\int_t^\infty du \int_{\mathbb{R}^d} d\xi \|\xi\|^{-\gamma} e^{-2u\|\xi\|^\alpha} \left|1 - e^{-i\langle\xi,x\rangle}\right|^2\right)$$

where we used the Parseval formula (6.3). Now, by using Fubini and computing the integral du, we get

$$\mathbf{E}|S_{\alpha,\delta}(x)|^2 = (2\pi)^{-d} \int_{\mathbb{R}^d} d\xi \|\xi\|^{-\gamma-\alpha}(1 - \cos(\xi \cdot x))e^{-2t\|\xi\|^\alpha} < \infty.$$

The function under the integral $d\xi$ is integrable at infinity because of the presence of the exponential function, while in the vicinity of zero we use $|1 - \cos(\langle\xi,x\rangle)| \le c\|\xi\|^2$ and then

$$\|\xi\|^{-\gamma-\alpha}(1 - \cos(\langle\xi,x\rangle))e^{-2t\|\xi\|^\alpha} \le C\|\xi\|^{-\alpha-\gamma+2}$$

which is integrable for ξ close to zero since $-\alpha - \gamma + 2 + d > 0$.

In the same we can control the increments of $S_{\alpha,\delta}$ to prove point 2. Indeed, for $x_1, x_2 \in \mathbb{R}^d$,

$$\mathbf{E}\left|S_{\alpha,\delta}(x_2) - S_{\alpha,\delta}(x_1)\right|^2$$

$$= \int_t^\infty du \int_{\mathbb{R}^d} d\xi \|\xi\|^{-\gamma} e^{-2u\|\xi\|^\alpha} \left|1 - e^{-i\langle\xi,(x_2-x_1)\rangle}\right|^2$$

$$2\int_t^\infty du \int_{\mathbb{R}^d} d\xi \|\xi\|^{-\gamma} e^{-2u\|\xi\|^\alpha} (1 - \cos(\langle\xi,(x_2-x_1)\rangle)$$

$$= \int_{\mathbb{R}^d} d\xi \|\xi\|^{-\gamma-\alpha}(1 - \cos(\langle\xi,(x_2-x_1)\rangle)))e^{-2t\|\xi\|^\alpha}$$

$$\le (\|x_2-x_1\|^2 \int_{\mathbb{R}^d} d\xi \|\xi\|^{-\gamma-\alpha+2} e^{-2t\|\xi\|^\alpha}$$

since $1 - \cos(\langle\xi,(x_2-x_1)\rangle)) \le \|x_2-x_1\|^2\|\xi\|^2$. For point 2., we refer to the proof of Proposition 6.4. $\square$

Let us now define

$$V_{\alpha,\delta}(t,x) = u(t,x) - S_\alpha, \delta(t,x).$$

We will show that $(U(x), x \in \mathbb{R}^d)$ is modulo a constant, an isotropic fBm.

Proposition 6.12. *Let $V_{\alpha,\gamma}$ as above. Then $(U_{\alpha,\gamma}(t,x), x \in \mathbb{R}^d)$ coincides in law with*

$$\left(\sqrt{C_{0,\alpha,\gamma}} B^{\frac{\alpha+\gamma-d}{2}}, x \in \mathbb{R}^d \right)$$

where $B^{\frac{\alpha+\gamma-d}{2}}$ is an isotropic fBm and $C^2_{0,\alpha,\gamma}$ given by

$$C_{0,\alpha,\gamma} = \int_{\mathbb{R}^d} dw \|w\|^{-(\alpha+\gamma)}(1 - \cos(\langle w, e_1 \rangle)) \text{ with } e_1 = (1, 0 \dots, 0). \quad (6.31)$$

Proof. We have for $x, y \in \mathbb{R}^d$,

$$V_{\alpha,\gamma}(t,x) - V_{\alpha,\gamma}(t,y)$$
$$= -\int_t^\infty \int_{\mathbb{R}^d} (G_\alpha(u, x-z) - G_\alpha(u, y-z)) \, W^\gamma(du, dz)$$
$$+ \int_0^t \int_{\mathbb{R}^d} (G_\alpha(u, x-z) - G_\alpha(u, y-z)) \, W^\gamma(du, dz).$$

We can write, for $x, y \in \mathbb{R}$, by using the independence of u and S (because the noise W^γ is white in time)

$$\mathbf{E} \left| V_{\alpha,\gamma}(t,x) - V_{\alpha,\gamma}(t,y) \right|^2$$
$$= \mathbf{E} \left[\int_0^t \int_{\mathbb{R}^d} (G_\alpha(u, x-z) - G_\alpha(u, y-z)) \, W^\gamma(du, dz) \right]^2$$
$$+ \mathbf{E} \left[\int_t^\infty \int_{\mathbb{R}^d} (G_\alpha(u, x-z) - G_\alpha(u, y-z)) \, W^\gamma(du, dz) \right]^2$$
$$= \mathbf{E} \left[\int_0^\infty \int_{\mathbb{R}^d} (G_\alpha(u, x-z) - G_\alpha(u, y-z)) \, W^\gamma(du, dz) \right]^2$$
$$= 2 \int_0^\infty du \int_{\mathbb{R}^d} d\xi \|\xi\|^{-\gamma} e^{-2u\|\xi\|^\alpha} (1 - \cos(\langle \xi, (x-y) \rangle))$$
$$= \int_{\mathbb{R}^d} d\xi \|\xi\|^{-\gamma-\alpha}(1 - \cos(\langle \xi, (x-y) \rangle))$$

where we computed the integral du. We will have

$$\mathbf{E} \left| V_{\alpha,\gamma}(x) - V_{\alpha,\gamma}(y) \right|^2 = \|y - x\|^{\alpha+\gamma-d} \int_{\mathbb{R}^d} dw \|w\|^{-(\alpha+\gamma)}(1 - \cos(\langle w, e_1 \rangle))$$

with $e_1 = (1, 0, \dots, 0)$. The last identity follows from Proposition 2 in [Herbin (2006)]. The last relation implies that $V_{\alpha,\gamma}$ coincides in law with $\sqrt{C_{0,\alpha,\gamma}} B^{\frac{\alpha+\gamma-d}{2}}$ where $B^{\frac{\alpha+\gamma-d}{2}}$ is an isotropic fBm and $C^2_{0,\alpha,\gamma}$ given by (6.31). $\qquad \square$

Theorem 6.2. *Fix $t > 0$. Then the process $(u(t,x), x \in \mathbb{R}^d)$ has the same finite dimensional distribution as*

$$\left(\sqrt{C_{0,\alpha,\gamma}} B^{\frac{\alpha+\gamma-d}{2}}(x) + S_{\alpha,\gamma}(x), x \in \mathbb{R}^d \right)$$

where $\left(B^{\frac{\alpha+\gamma-d}{2}}(x), x \in \mathbb{R}^d \right)$ is an isotropic multiparameter fBm with Hust index $\frac{\alpha+\gamma-d}{2}$ and $\left(S_{\alpha,\gamma}(x), x \in \mathbb{R}^d \right)$ is a Gaussian process with C^∞ sample paths.

Chapter 7

The stochastic wave equation with space-time white noise

The (homogenous) wave equation is an important second-order partial differential equation. It modelizes the vibrations of a perfectly flexible string. The simplest way to express this equation is

$$\frac{\partial^2 u}{\partial t^2}(t, x) = \Delta u(t, x), \quad t \geq 0, x \in \mathbb{R}$$

where Δ is the standard Laplacian on $\mathbb{R}$. The above equation corresponds to the case of an infinite string, since we assumed that the space variable x belongs to the whole real line. In the case of a finite string, one usually supposes that $x \in [0, L] \subset \mathbb{R}$ with $L > 0$ being the length of the string. $u(t, x)$ constitutes the position at time t of the point x on the string. The inhomogenous wave equation is written as

$$\frac{\partial^2 u}{\partial t^2}(t, x) = \Delta u(t, x) + f(t, x), \quad t \geq 0, x \in \mathbb{R}$$

f being called the source function and it usually represents the force driving the wave on the string. When this force depends on random factors, then we deal with the stochastic wave equation. As for the heat equation, the most common way to model the random source is by the space-time white noise, but a correlation in time and/or in space can be also considered.

7.1 General properties

Consider the linear stochastic wave equation driven by a white-colored noise W. That is,

$$\frac{\partial^2 u}{\partial t^2}(t, x) = \Delta u(t, x) + \dot{W}(t, x), \quad t \geq 0, x \in \mathbb{R}^d \tag{7.1}$$

$$u(0, x) = 0, \quad x \in \mathbb{R}^d$$

$$\frac{\partial u}{\partial t}(0, x) = 0, \quad x \in \mathbb{R}^d.$$

Here Δ is the Laplacian on $\mathbb{R}^d$, see (5.2), and

$$W = \big(W(t, A); t \geq 0, A \in \mathcal{B}_b(\mathbb{R}^d)\big)$$

is a space-time white noise, i.e. a centered Gaussian field with covariance (5.4) (see Section 3.2 for its definition and properties). The values of $u(0, x)$ and $\frac{\partial u}{\partial t}(0, x)$ represents the initial values (the position and the speed at time $t = 0$ of the point x on the string) but it can be also supposed to be nonzero.

A key factor for the analysis of the solution to the wave equation (7.1) is the fundamental solution or the Green kernel.

Let us denote by G_1 the fundamental solution, i.e. the function which satisfies

$$\frac{\partial^2 u}{\partial t^2} - \Delta u = 0.$$

It is known that G_1 has the following expression in spatial dimensions $d = 1, 2, 3$ (see e.g. [Treves (1975)]).

$$G_1(t, x) = \frac{1}{2} 1_{\{|x| < t\}}, \quad \text{if } d = 1 \tag{7.2}$$

$$G_1(t, x) = \frac{1}{2\pi} \frac{1}{\sqrt{t^2 - \|x\|^2}} 1_{\{\|x\| < t\}}, \quad \text{if } d = 2$$

$$G_1(t, x) = c_d \frac{1}{t} \sigma_t, \quad \text{if } d = 3,$$

where σ_t denotes the surface measure on the 3-dimensional sphere of radius t. The easiest way to define G_1 for all $d \geq 1$ is via its Fourier transform

$$\mathcal{F}G_1(t, \cdot)(\xi) = \frac{\sin(t\|\xi\|)}{\|\xi\|}, \tag{7.3}$$

for any $\xi \in \mathbb{R}^d, t > 0, d \geq 1$. The formula (7.3) is the same for all dimension $d \geq 1$.

To motivate the definition of the solution (7.4), let us make the following remark.

Remark 7.1. Let G_1 be the Green kernel (7.2). For every $\varphi \in C^2([0, \infty) \times \mathbb{R}^d)$, define

$$v(t, x) = \int_0^t ds \int_{\mathbb{R}^d} dy \, G_1(t - s, x - y)\varphi(s, y), \quad t \geq 0, x \in \mathbb{R}^d.$$

Then v satisfies the ordinary differential equation

$$\frac{\partial^2 v}{\partial t^2}(t, x) = \Delta v(t, x) + \varphi(t, x), \quad t \geq 0, x \in \mathbb{R}^d$$

As for the heat equation, the solution to the stochastic wave equation (7.1) will be defined by replacing $\dot{W}(s, y)dsdy$ by the stochastic integral $W(ds, dy)$. See also Remark 5.1 for a similar result for the heat equation.

The mild (or evolutive) solution of (7.1) is a square-integrable process $u = \big(u(t,x); t \geq 0, x \in \mathbb{R}^d\big)$ defined by the Wiener integral representation with respect to the space-time white noise

$$u(t,x) = \int_0^t \int_{\mathbb{R}^d} G_1(t-s, x-y) W(ds, dy), \quad t \geq 0, x \in \mathbb{R}. \tag{7.4}$$

The above Wiener integral is that described in Section 4.4. Another way the define the solution to (7.1) is in the weak sense. The relationship between the two concepts is discussed in Section 7.1.1.

As for the heat equation, let us start by giving a necessary and sufficient condition for the existence of the mild solution.

Proposition 7.1. *The mild solution (7.4) is well-defined if and only if* $d = 1$. *If* $d = 1$, *then for every* $t > 0, x \in \mathbb{R}$,

$$u(t,x) \sim N\left(0, K_{2,t}\right) \tag{7.5}$$

with $K_{2,t} = \frac{t^2}{4} < \infty$. *Moreover for* $d = 1$ *and for every* $T > 0$ *and* $p \geq 1$,

$$\sup_{t \in [0,T], x \in \mathbb{R}} \mathbf{E}|u(t,x)|^p \leq C_T \tag{7.6}$$

with $C_T > 0$.

Proof. The mild solution is well-defined if and only if $\mathbf{E}u(t,x)^2 < \infty$ for every $t \geq 0$ and $x \in \mathbb{R}^d$. Via the expression of the Fourier transform of the Green kernel (see (7.3)) and by applying as usually the Parseval relation, we can write

$$\mathbf{E}u(t,x)^2 = \int_0^t \int_{\mathbb{R}^d} G_1(t-s, x-y)^2 \, dy \, ds$$

$$= (2\pi)^{-d} \int_0^t ds \int_{\mathbb{R}^d} d\xi \, \mathcal{F}G_1(t-s, x-\cdot)(\xi) \overline{\mathcal{F}G_1(t-s, x-\cdot)(\xi)}$$

$$= (2\pi)^{-d} \int_0^t ds \int_{\mathbb{R}^d} d\xi \frac{\sin^2(s\|\xi\|)}{\|\xi\|^2}. \tag{7.7}$$

Let us prove the following inequality, for every $t \geq 0, \xi \in \mathbb{R}^d$,

$$C_{1,t} \frac{1}{1 + \|\xi\|^2} \leq \int_0^t ds \frac{\sin^2(s\|\xi\|)}{\|\xi\|^2} \leq C_{2,t} \frac{1}{1 + \|\xi\|^2} \tag{7.8}$$

with $0 < C_{1,t} < C_{2,t}$ (these constants may depend on t). Denote

$$I(t,\xi) = \int_0^t ds \frac{\sin^2(s\|\xi\|)}{\|\xi\|^2}.$$

Upper bound. Assume first $\|\xi\| \leq 1$. Then using $\sin^2(s\|\xi\|) \leq s^2\|\xi\|^2$, we can write (below the various constants may change from one line to another or even on the same line!)

$$I(t,\xi) \leq \int_0^t ds \times s^2 = \frac{1}{3}t^3 \leq C'_{2,t}\frac{1}{1+\|\xi\|^2}$$

since $\frac{1}{1+\|\xi\|^2} \geq \frac{1}{2}$ if $\|\xi\| \leq 1$, with $C'_{2,t} = \frac{2}{3}t^3$.

Now, let $\|\xi\| > 1$. We can write, via the change of variables $\tilde{s} = s\|\xi\|$,

$$I(t,\xi) = \frac{1}{\|\xi\|^3}\int_0^{t\|\xi\|} ds\,\sin^2(s)ds = \frac{1}{2\|\xi\|^3}\int_0^{t\|\xi\|}(1-\cos(2s))ds$$

$$= \frac{1}{2\|\xi\|^3}\left(t\|\xi\| - \frac{1}{2}\sin(2t\|\xi\|)\right)$$

and thus

$$I(t,\xi) \leq \frac{1}{2\|\xi\|^3}(t\|\xi\| + \frac{1}{2}) \leq \frac{1}{2\|\xi\|^2}\left(t + \frac{1}{2}\right)$$

$$\leq \left(t + \frac{1}{2}\right)\frac{1}{1+\|\xi\|^2} = C''_{2,t}\frac{1}{1+\|\xi\|^2}$$

where we used $\frac{1}{2\|\xi\|^2} \leq \frac{1}{1+\|\xi\|^2}$ for $\|\xi\| > 1$. Thus the upper bound holds with $C_{2,t} = C'_{2,t} + C''_{2,t} = \frac{2}{3}t^3 + \left(t + \frac{1}{2}\right)$.

Lower bound. We can assume $t = 1$. Also assume $\|\xi\| \leq 1$. Then

$$I(1,\xi) = \frac{1}{2\|\xi\|^2}\int_0^1 (1-\cos(2s\|\xi\|))ds$$

$$= \frac{1}{2\|\xi\|^2}\left(1 - \frac{1}{2\|\xi\|}\sin(2\|\xi\|)\right)$$

$$\geq C_1\frac{1}{\|\xi\|^2} \geq C_t\frac{1}{1+\|\xi\|^2}.$$

If $\|\xi\| > 1$, since $\frac{1}{1+\|\xi\|^2} < \frac{1}{|\xi|^2}$,

$$I(1,\xi) \geq \frac{1}{1+\|\xi\|^2}\frac{1}{2}\int_0^{1t}(1-\cos(2s\|\xi\|))ds$$

$$= \frac{1}{2}\frac{1}{1+|\xi|^2}\left(1 - \frac{\sin(2\|\xi\|)}{2\|\xi\|}\right)$$

We use the inequality $\frac{\sin(2\|\xi\|)}{2\|\xi\|} \leq \frac{1}{2}$ for $\|\xi\| > 1$ and we get

$$I(1,\xi) \geq \frac{1}{4}\frac{1}{1+\|\xi\|^2}.$$

Now, the integral $\int_{\mathbb{R}^d} \frac{1}{1+\|x\|^2} d\xi$ is finite if and only if $d = 1$ and thus the double inequality (7.8) gives the conclusion. Clearly, with $C_{2,t} = \frac{2}{3}t^3 + (t + \frac{1}{2})$, we have for $t \in [0, T]$ and $x \in \mathbb{R}$

$$\mathbf{E}u(t, x)^2 \leq C_{2,T} \int_{\mathbb{R}} \frac{1}{1 + \|x\|^2} dx \leq C_T$$

and this implies (7.6).

If $d = 1$, we have from (7.7), since $\int_{\mathbb{R}} \frac{(\sin \|\xi\|)^2}{\|\xi\|^2} d\xi = \pi$,

$$\mathbf{E}u(t, x)^2 = \frac{t^2}{4\pi} \int_{\mathbb{R}} \frac{(\sin \|\xi\|)^2}{\|\xi\|^2} d\xi = \frac{t^2}{4} = K_{2,t}$$

and this gives (7.5). The bound (7.6) is obvious from (7.5) due to the fact that $(u(t, x), t \geq 0)$ is a Gaussian process.

$\square$

Remark 7.2.

One can show by a direct proof that for $d = 2$ the mild solution is not well-defined. If $t > 0$ and $x \in \mathbb{R}^2$ are fixed in $\mathbb{R}_+$, then the function

$$s \to G_1(t - s, x - y)1_{[0,t]}(s) : \mathbb{R}_+ \to \mathbb{R}$$

is not square-integrable. Indeed, from (7.2),

$$\int_0^t \int_{\mathbb{R}^2} G_1^2(t - s, x - y) dy ds$$

$$= \frac{1}{4\pi^2} \int_0^t \int_{\mathbb{R}^2} \frac{1}{(t - s)^2 - \|x - y\|^2} 1_{\|x-y\|<t-s} dy dx$$

$$= \frac{1}{4\pi^2} \int_0^t \int_{\|x-y\|\leq t-s} \frac{1}{(t - s)^2 - \|x - y\|^2} dy ds$$

$$= \frac{1}{4\pi^2} \int_0^t \int_{(\|v\|\leq u)\subset\mathbb{R}^2} \frac{1}{u^2 - \|v\|^2} dv du$$

and by the change of variables in polar coordinates $v = (r \cos(\theta), r \sin(\theta))$ with $r > 0, \theta \in [0, 2\pi)$, we obtain

$$\int_0^t \int_{\mathbb{R}^2} G_1^2(t - s, x - y) dy ds$$

$$= \frac{1}{4\pi^2} \int_0^t ds \int_0^{2\pi} d\theta \int_0^u \frac{r}{u^2 - r^2}$$

$$= \frac{1}{4\pi} \int_0^t du \int_0^u \frac{2r}{u^2 - r^2}$$

$$= \frac{1}{4\pi} \int_0^t du \left(-\log(u^2 - r^2)\Big|_{r-0}^{r=u} \right) = \infty.$$

Let us compute the joint covariance of the solution (with respect to both its time and space variables).

Lemma 7.1. *Let u be given by (7.4). For every $t_1, t_2 > 0$ and $x, y \in \mathbb{R}$, we have*

$$\mathbf{E}\big[u(t_1, x)u(t_2, y)\big] = \frac{1}{16}1_{\{|t_1-t_2|\leq|y-x|<t_1+t_2\}}\,(t_1 + t_2 - |x - y|)^2$$

$$+\frac{1}{4}1_{\{|t_1-t_2|>|y-x|\}}(t_1 \wedge t_2)^2. \tag{7.9}$$

Proof. By the isometry of the Wiener integral and from (7.2), we obtain

$$\mathbf{E}\big[u(t_1, x)u(t_2, y)\big]$$

$$= \frac{1}{4}\int_0^{t_1\wedge t_2} ds \int_{\mathbb{R}} dz\, G_1(t_1 - u, x - z)G_1(t_2 - u, y - z)$$

$$= \frac{1}{4}\int_0^{t_1\wedge t_2} ds \int_{\mathbb{R}} dz\, 1_{\{|x-z|\leq t_1-s\}}1_{\{|y-z|\leq t_2-s\}}$$

$$= \frac{1}{4}\int_0^{t_1\wedge t_2 \wedge \frac{1}{2}(t_1+t_2-|y-x|)} ds \int_{\mathbb{R}} dz\, 1_{\{|x-z|\leq t_1-s\}}1_{\{|y-z|\leq t_2-s\}}$$

$$+\frac{1}{4}\int_0^{t_1\wedge t_2} ds \int_{\mathbb{R}} dz\, 1_{\{2s>t_1+t_2-|y-x|\}}1_{\{|x-z|\leq t_1-s\}}1_{\{|y-z|\leq t_2-s\}}$$

$$= 1_{\{t_1+t_2>|y-x|\}}\frac{1}{4}\int_0^{t_1\wedge t_2 \wedge \frac{1}{2}(t_1+t_2-|y-x|)} ds \int_{\mathbb{R}} dz\, 1_{\{|x-z|\leq t_1-s\}}1_{\{|y-z|\leq t_2-s\}}$$

$$= 1_{\{t_1+t_2>|y-x|\}}\frac{1}{4}\int_0^{t_1\wedge t_2 \wedge \frac{1}{2}(t_1+t_2-|y-x|)} ds \left(\int_{(x-t_1+s)\vee(y-t_2+s)}^{(x+t_1-s)\vee(y+t_2-s)} dz\right)_+.$$

In order to find the integration domain for the integral dz, we will consider several situations. Assume $x \geq y$.

If $t_1 \geq t_2$ and $x - y \geq t_1 - t_2$ then

$$y + t_2 - s \leq x + t_1 - s \quad \text{and} \quad x - t_1 + s \geq y - t_2 + s.$$

In this case,

$$\mathbf{E}\big[u(t_1,x)u(t_2,y)\big]$$

$$= 1_{\{t_1+t_2>|y-x|\}}\frac{1}{4}\int_0^{t_1\wedge t_2\wedge\frac{1}{2}(t_1+t_2-|y-x|)} ds\left(\int_{x-t_1+s}^{y+t_2-s} dz\right)_+$$

$$= 1_{\{t_1+t_2>|y-x|\}}\frac{1}{4}\int_0^{t_1\wedge t_2\wedge\frac{1}{2}(t_1+t_2-|y-x|)} ds(t_1+t_2-(x-y)-2s)$$

$$= 1_{\{t_1+t_2>|y-x|\}}\frac{1}{4}\int_0^{\frac{1}{2}(t_1+t_2-|y-x|)} ds(t_1+t_2-(x-y)-2s)$$

$$= 1_{\{t_1+t_2>|y-x|\}}\frac{1}{16}(t_1+t_2-(x-y))^2.$$

If $t_1 \geq t_2$ and $x-y < t_1-t_2$ then

$$y+t_2-s \leq x+t_1-s \text{ and } x-t_1+s \leq y-t_2+s.$$

Then

$$\mathbf{E}\big[u(t_1,x)u(t_2,y)\big]$$

$$= 1_{\{t_1+t_2>|y-x|\}}\frac{1}{4}\int_0^{t_1\wedge t_2\wedge\frac{1}{2}(t_1+t_2-|y-x|)} ds\int_{y-t_2+s}^{y+t_2-s} dz$$

$$= 1_{\{t_1+t_2>|y-x|\}}\frac{1}{4}\int_0^{t_1\wedge t_2\wedge\frac{1}{2}(t_1+t_2-|y-x|)} ds(2t_2-2s)$$

$$= 1_{\{t_1+t_2>|y-x|\}}\frac{1}{4}\int_0^{t_2} ds(2t_2-2s) = 1_{\{t_1+t_2>|y-x|\}}\frac{1}{4}t_2^2.$$

If $t_1 \leq t_2$ and $x-y \leq t_2-t_1$, then

$$x+t_1-s \leq y+t_2-s \text{ and } x-t_1+s \geq y-t_2+s.$$

So

$$\mathbf{E}\big[u(t_1,x)u(t_2,y)\big]$$

$$= 1_{\{t_1+t_2>|y-x|\}}\frac{1}{4}\int_0^{t_1\wedge t_2\wedge\frac{1}{2}(t_1+t_2-|y-x|)} ds\int_{x-t_1+s}^{x+t_1-s} dz$$

$$= 1_{\{t_1+t_2>|y-x|\}}\frac{1}{4}\int_0^{t_1\wedge t_2\wedge\frac{1}{2}(t_1+t_2-|y-x|)} ds(2t_1-2s)$$

$$= 1_{\{t_1+t_2>|y-x|\}}\frac{1}{4}\int_0^{t_1} ds(2t_1-2s) = 1_{\{t_1+t_2>|y-x|\}}\frac{1}{4}t_1^2.$$

If $t_1 \leq t_2$ **and** $x - y \geq t_2 - t_1$, **then**

$$y + t_2 - s \leq x + t_1 - s \text{ and } x - t_1 + s \geq y - t_2 + s$$

Consequently

$$
\mathbf{E}\big[u(t_1, x)u(t_2, y)\big]
$$

$$
= 1_{\{t_1+t_2 > |y-x|\}} \frac{1}{4} \int_0^{t_1 \wedge t_2 \wedge \frac{1}{2}(t_1+t_2-|y-x|)} ds \left(\int_{x-t_1+s}^{y+t_2-s} dz \right)_+
$$

$$
= 1_{\{t_1+t_2 > |y-x|\}} \frac{1}{4} \int_0^{\frac{1}{2}(t_1+t_2-|y-x|)} ds(t_1 + t_2 - (x-y) - 2s)
$$

$$
= 1_{\{t_1+t_2 > |y-x|\}} \frac{1}{16} (t_1 + t_2 - (x-y))^2.
$$

By the above four estimates, we showed (7.9) when $x \geq y$. By symmetry, it is also valid for $x < y$. $\qquad\square$

The joint covariance formula (7.9) is pretty complex. Simpler expressions will be obtained when we fix the temporal or the spatial variable.

7.1.1   *Relation with the weak solution*

Let us give the meaning the weak solution to the stochastic wave equation.

Definition 7.1. We will say that the random field $(u(t, x), t \geq 0, x \in \mathbb{R})$ is a weak solution to the stochastic wave equation (7.1) if for every $T > 0$

$$
\int_0^T \int_{\mathbb{R}} u(t, x) \left(\frac{\partial^2 \varphi}{\partial t^2}(t, x) - \frac{\partial^2 \varphi}{\partial x^2}(t, x) \right) dx dt = \int_0^T \int_{\mathbb{R}} \varphi(s, y) W(ds, dy)
$$

$$(7.10)$$

for every $\varphi \in C^\infty(\mathbb{R}_+ \times \mathbb{R})$ with compact support such that

$$
\varphi(T, x) = 0 \text{ and } \frac{\partial \varphi}{\partial x}(T, x) = 0 \text{ for every } x \in \mathbb{R}. \qquad (7.11)
$$

Proposition 7.2. *A random field* $(u(t, x), t \geq 0, x \in \mathbb{R})$ *is a mild solution to (7.1) if and only if it is a weak solution to (7.1).*

Proof. Assume u is given by (7.4). Take $\varphi \in C^\infty(\mathbb{R}+ \times \mathbb{R})$ satisfying

(7.11). We can write, for ant $T > 0$,

$$\int_0^T \int_{\mathbb{R}} u(t, x) \left(\frac{\partial^2 \varphi}{\partial t^2}(t, x) - \frac{\partial^2 \varphi}{\partial x^2}(t, x) \right) dx dt$$

$$= \int_0^T dt \int_{\mathbb{R}} dx \left(\int_0^t \int_{\mathbb{R}} G_1(t - u, x - y) W(ds, dy) \right)$$

$$\times \left(\frac{\partial^2 \varphi}{\partial t^2}(t, x) - \frac{\partial^2 \varphi}{\partial x^2}(t, x) \right)$$

$$= \int_0^T \int_{\mathbb{R}} W(ds, dy) \left[\int_s^T dt \int_{\mathbb{R}} dx G_1(t - s, x - y) \right.$$

$$\left. \times \left(\frac{\partial^2 \varphi}{\partial t^2}(t, x) - \frac{\partial^2 \varphi}{\partial x^2}(t, x) \right) \right].$$

Let us compute the integral $dt dx$, i.e.

$$\int_s^T dt \int_{\mathbb{R}} dx G_1(t - s, x - y) \left(\frac{\partial^2 \varphi}{\partial t^2}(t, x) - \frac{\partial^2 \varphi}{\partial x^2}(t, x) \right)$$

$$= \frac{1}{2} \int_s^T dt \int_{\mathbb{R}} dx 1_{|x-y|<t-u} \left(\frac{\partial^2 \varphi}{\partial t^2}(t, x) - \frac{\partial^2 \varphi}{\partial x^2}(t, x) \right).$$

First,

$$\frac{1}{2} \int_s^T dt \int_{\mathbb{R}} dx 1_{|x-y|<t-u} \frac{\partial^2 \varphi}{\partial t^2}(t, x)$$

$$= \frac{1}{2} \int_s^T dt \int_{y-(t-s)}^{y+(t-s)} dx \frac{\partial^2 \varphi}{\partial t^2}(t, x)$$

$$= \frac{1}{2} \int_{y-(T-s)}^{y+(T-s)} dx \int_s^T dt 1_{t>x-y+s} 1_{t>y-x+s} \frac{\partial^2 \varphi}{\partial t^2}(t, x)$$

$$= \frac{1}{2} \int_{y-(T-s)}^{y} dx \int_{y-x+s}^T dt \frac{\partial^2 \varphi}{\partial t^2}(t, x)$$

$$+ \frac{1}{2} \int_y^{y+(T-s)} dx \int_{x-y+s}^T dx \frac{\partial^2 \varphi}{\partial t^2}(t, x).$$

By using the integration by parts and assumption (7.11),

$$\frac{1}{2} \int_s^T dt \int_{\mathbb{R}} dx 1_{|x-y|<t-u} \frac{\partial^2 \varphi}{\partial t^2}(t, x)$$

$$= -\frac{1}{2} \int_{y-(T-s)}^{y} dx \partial_1 \varphi(y - x + s, x) dx$$

$$- \frac{1}{2} \int_y^{y+(T-s)} \partial_1 \varphi(x - y + s, x) dx \qquad (7.12)$$

where $\partial_1\varphi$ means the derivative of φ with respect to its first variable. On the other hand,

$$\frac{1}{2}\int_s^T dt \int_{\mathbb{R}} dx 1_{|x-y|<t-u}\frac{\partial^2\varphi}{\partial x^2}(t,x)$$

$$=\frac{1}{2}\int_s^T dt \int_{y-(t-s)}^{y+(t-s)} \frac{\partial^2\varphi}{\partial x^2}(t,x)$$

$$=\frac{1}{2}\int_s^T dt\partial_2\left(\varphi(t,y+t-s)-\partial_2\varphi(t,y-(t-s))\right)$$

where $\partial_2\varphi$ is the partial derivative of φ with respect to its second variable. So

$$\frac{1}{2}\int_s^T dt \int_{\mathbb{R}} dx 1_{|x-y|<t-u}\frac{\partial^2\varphi}{\partial x^2}(t,x)$$

$$=\frac{1}{2}\int_y^{y+T-s} dx\partial_2\varphi(s-y+x,x)dx$$

$$-\frac{1}{2}\int_{y-(T-s)}^{y} dx\partial_2\varphi(s+y-x,x)dy. \tag{7.13}$$

By putting together (7.12) and (7.13), we will have

$$\int_s^T dt \int_{\mathbb{R}} dxG_1(t-s,x-y)\left(\frac{\partial^2\varphi}{\partial t^2}(t,x)-\frac{\partial^2\varphi}{\partial x^2}(t,x)\right)$$

$$=-\frac{1}{2}\int_y^{y+T-s}\left(\partial_1\varphi(s-y+x,x)+\partial_2\varphi(s-y+x,x)\right)ds$$

$$+\frac{1}{2}\int_{y-(T-s)}^{y}\left(\partial_2\varphi(s+y-x,x)-\partial_1\varphi(s+y-x,x)\right)dx$$

and this can be written as

$$\int_s^T dt \int_{\mathbb{R}} dxG_1(t-s,x-y)\left(\frac{\partial^2\varphi}{\partial t^2}(t,x)-\frac{\partial^2\varphi}{\partial x^2}(t,x)\right)$$

$$=\frac{1}{2}\int_y^{y+T-s} -\frac{d}{dx}\varphi(s-y+x,x)dx+\frac{1}{2}\int_{y-(T-s)}^{y} \frac{d}{dx}\varphi(s+y-x,x)dx$$

$$=\frac{1}{2}\left(-\varphi(T,y+T-s)+2\varphi(s,y)-\varphi(T,y-T+s)\right)$$

$$=\varphi(s,y).$$

Consequently, for every $T > 0$,

$$\int_0^T \int_{\mathbb{R}} u(t,x) \left(\frac{\partial^2 \varphi}{\partial t^2}(t,x) - \frac{\partial^2 \varphi}{\partial x^2}(t,x) \right) dx\,dt$$

$$= \int_0^T \int_{\mathbb{R}} W(ds,dy) \left[\int_s^T dt \int_{\mathbb{R}} dx\, G_1(t-s, x-y) \left(\frac{\partial^2 \varphi}{\partial t^2}(t,x) - \frac{\partial^2 \varphi}{\partial x^2}(t,x) \right) \right]$$

$$= \int_0^T \int_{\mathbb{R}} \varphi(s,y) W(ds,dy),$$

so u satisfies the definition of the weak solution.

Assume now u is a weak solution. We define, for every $\varepsilon > 0$ and $(s,y) \in [0,T] \times \mathbb{R}$,

$$\varphi_\varepsilon(s,y) = \int_s^T \int_{\mathbb{R}} \Phi_\varepsilon(t-s, x-y) G_1(-s, x-y) dy\,ds$$

where Φ_ε is a family of functions in $C^\infty(\mathbb{R}_+ \times \mathbb{R})$ with support included in $\{x \in \mathbb{R}^2, |x| \leq 1\}$ such that

$$\int_{\mathbb{R}^2} \Phi_\varepsilon(s,y) dy\,ds = 1.$$

We can see that for every $\varepsilon > 0$, φ_ε belongs to $C^\infty([0,T] \times \mathbb{R})$ and it has compact support. Moreover,

$$\varphi_\varepsilon(T,y) = \partial_1 \varphi_\varepsilon(T,y) = 0$$

and

$$\partial_1^2 \varphi_\varepsilon(s,y) - \partial_2^2 \varphi_\varepsilon(s,y) = \Phi_\varepsilon(t-s, x-y)$$

for $s \leq t$ and $y \in \mathbb{R}$ (we denoted $\partial_1 \varphi_\varepsilon(s,y)$ the partial derivative of φ_ε with respect s and $\partial_1^2 \varphi_\varepsilon(s,y)$ is the second derivative of φ_ε with respect s; similarly ∂_2 and ∂_2^2 stand for the similar derivative with respect to the space variable). Thus we can apply (7.10) to φ_ε to write

$$\int_0^T \int_{\mathbb{R}} u(t,x) \left(\frac{\partial^2 \varphi_\varepsilon}{\partial t^2}(t,x) - \frac{\partial^2 \varphi_\varepsilon}{\partial x^2}(t,x) \right) dx\,dt = \int_0^T \int_{\mathbb{R}} \varphi_\varepsilon(s,y) W(ds,dy)$$

$$(7.14)$$

Take O an open set contained in $[0,T] \times \mathbb{R}$. Then for every $(t,x) \in O$,

$$\int_0^T \int_{\mathbb{R}} u(s,y) \left(\partial_1^2 \varphi_\varepsilon(s,y) - \partial_2^2 \varphi_\varepsilon(s,y) \right) dy\,ds$$

$$= \int_0^T \int_{\mathbb{R}} u(s,y) \Phi_\varepsilon(t-s, x-y) dy\,ds = (u * \phi_\varepsilon)(t,x)$$

$$\to_{\varepsilon \to 0} u(t,x) \text{ in } L^2(O) \text{ almost surely,} \qquad (7.15)$$

where "$*$" stands for the convolution product. Now, define the compact set

$$I := \bigcup_{y \in \mathbb{R}} \{supp \, \varphi_\varepsilon(t, y), (t, y) \in O\}.$$

We can show that pour tout $(t, x) \in I$, we have the convergence

$$\varphi_\varepsilon(s, y) \to_{\varepsilon \to 0} G_1(t - s, x - y) \text{ in } L^2([0, T] \times I).$$

By the Wiener isometry,

$$\mathbf{E} \left(\int_0^T \int_{\mathbb{R}} (\varphi_\varepsilon(s, y) - G_1(t - s, x - y)) \, W(ds, dy) \right)^2$$

$$= \mathbf{E} \int_0^T \int_{\mathbb{R}} (\varphi_\varepsilon(s, y) - G_1(t - s, x - y))^2 \, dy ds \to_{\varepsilon \to 0} 0$$

for every $(t, x) \in [0, T] \times \mathbb{R}$. Thus, for $(t, x) \in [0, T] \times \mathbb{R}$,

$$\int_0^T \int_{\mathbb{R}} \varphi_\varepsilon(s, y) W(ds, dy) \to_{\varepsilon \to 0} \int_0^T \int_{\mathbb{R}} G_1(t - s, x - y) W(ds, dy) \text{ in } L^2(\Omega)$$

(7.16)

and this implies the almost sure convergence for a subsequence. By taking the limit as ε in (7.14), via (7.15) and (7.16), we get almost surely for every $(t, x) \in [0, T] \times \mathbb{R}$

$$u(t, x) = \int_0^t \int_{\mathbb{R}} G_1(t - s, x - y) W(ds, dy).$$

$\square$

7.2　Behavior of the solution in time

We deal now with the Gaussian stochastic process $(u(t, x), t \geq 0)$, the spatial variable $x \in \mathbb{R}$ being fixed. The first step is to calculate the spatial covariance of the solution, by applying the covariance formula (7.9).

Proposition 7.3. *Let u be given by (7.4). For every $s, t \geq 0$ and for every $x \in \mathbb{R}$, we have*

$$\mathbf{E} u(t, x) u(s, x) = \frac{1}{4} (t \wedge s)^2. \tag{7.17}$$

Proof. It suffices to take $x = y$ in the covariance formula (7.9). $\square$

Remark 7.3. For $t = s$, we have

$$\mathbf{E} u(t, x)^2 = \frac{t^2}{4}$$

so we retrieve the variance of the solution given in (7.5).

We have the following link between the solution to the wave equation with space-time white noise and the Wiener process.

Corollary 7.1. *Let* u *be given by (7.4). For fixed* $x \in \mathbb{R}$, *the process* $(u(t,x), t \geq 0)$ *has the same finite dimensional distributions as*

$$\left(\frac{1}{2} W_{t^2}, t \geq 0 \right)$$

where $(W_t, t \geq 0)$ *is a standard Wiener process.*

Proof. Obviously, for $s, t \geq 0$,

$$\mathbf{E} \left(\frac{1}{2} W_{t^2} \frac{1}{2} W_{s^2} \right) = \frac{1}{4} (t^2 \wedge s^2) = \frac{1}{4} (t \wedge s)^2.$$

$\square$

We give below other distributional and trajectorial properties of the random field u from (7.4) with respect to the temporal variable.

Proposition 7.4. *Let* u *be given by (7.4) and let* $x \in \mathbb{R}$ *be fixed. Then we have the following.*

(1) The process $(u(t,x), t \geq 0)$ *is self-similar of index 1.*
(2) For every $s, t \geq 0$,

$$\mathbf{E} |u(t,x) - u(s,x)|^2 = \frac{1}{4} ((t \vee s)^2 - (t \wedge s)^2). \tag{7.18}$$

(3) Let $T > 0$ *arbitrary. For every* $s, t \in [0, T]$, *we have*

$$\mathbf{E} |u(t,x) - u(s,x)|^2 \leq C_2 |t - s|$$

with $C_2 = C_2(T) > 0$. *Consequently the sample path* $t \to u(t,x)$ *is almost surely Hölder continuous of order* δ *for every* $\delta \in \left(0, \frac{1}{2} \right)$.
(4) For every $t \geq s \geq t_0 > 0$, *we have*

$$\mathbf{E} |u(t,x) - u(s,x)|^2 \geq C_1 (t - s)$$

with $C_1 = C_1(t_0) > 0$.

Proof. Concerning the self-similarity in time, just notice that for every $c > 0$,

$$\mathbf{E} u(ct,x) u(cs,x) = \frac{1}{4} ((ct) \wedge (cs))^2 = \frac{1}{4} c^2 (t \wedge s)^2$$
$$= \mathbf{E} (cu(t,x) cu(s,x)).$$

For point 2., we have for every $t \geq s \geq 0$, via (7.17),

$$\mathbf{E} |u(t, x) - u(s, x)|^2 = \frac{1}{4}(t^2 + s^2 - 2(t \wedge s)^2) = \frac{1}{4}(t^2 - s^2)$$

$$= \frac{1}{4}((t \vee s)^2 - (t \wedge s)^2).$$

Thus, for $s, t \in [0, T]$ with $s \leq t$,

$$\mathbf{E} |u(t, x) - u(s, x)|^2 \leq \frac{1}{4} 2T(t - s) = C_2(T)(t - s)$$

and this is point 3. Finally, for $0 < t_0 \leq s \leq t$,

$$\mathbf{E} |u(t, x) - u(s, x)|^2 = \frac{1}{4}(t - s)(t + s) \geq \frac{t_0}{2}(t - s).$$

$\square$

We also have an interesting property concerning the independence of the increments in time of the random field u.

Proposition 7.5. *Let u be defined by (7.4). Let $0 \leq a < b \leq s < t$. Then*

$$\mathbf{E} (u(t, x) - u(s, x))(u(b, x) - u(a, x)) = 0.$$

Consequently, the temporal increments $u(t, x) - u(s, x)$ and $u(b, x) - u(a, x)$ are independent random variables.

Proof. By the covariance formula (7.17),

$$\mathbf{E} (u(t, x) - u(s, x))(u(b, x) - u(a, x))$$

$$= \frac{1}{4} ((t \wedge b)^2 - (t \wedge a)^2 - (s \wedge b)^2 + (s \wedge a)^2)$$

$$= \frac{1}{4}(b^2 - a^2 - b^2 + a^2) = 0.$$

Since the process $(u(t, x), t \geq 0)$ is Gaussian, the uncorrelation of the increments $u(t, x) - u(s, x)$ and $u(b, x) - u(a, x)$ implies their independence. $\square$

Remark 7.4.

- We notice from Proposition 7.4 that the paths $t \to u(t, x)$ is $\frac{1}{2}$-Hölder continuous and this order is sharp, i.e. this path cannot be δ-Hölder with $\delta > \frac{1}{2}$ (see Remark 5.3).
- We notice that for the process $(u(t, x), t \geq 0)$, its self-similarity index is 1 while its (sharp) Hölder continuity index is $\frac{1}{2}$, so these two indices do not coincide (for the fractional or bifractional Brownian motion or for the solution to the stochastic heat equation, with white or colored noise in space, the self-similarity order is the same as the Hölder continuity order).
- The stochastic wave with spatial variable in a finite interval $[0, L]$ has been studied in [Cabana (1970)].

7.3 Behavior of the solution in space

Fix $t \geq 0$ and consider the Gaussian random field $(u(t,x), x \in \mathbb{R})$. Its covariance reads as follows.

Proposition 7.6. *Let $(u(t,x), t \geq 0, x \in \mathbb{R})$ defined by (7.4). For every $t \geq 0$ and $x, y \in \mathbb{R}$, it holds that*

$$\mathbf{E}\Big(u(t,x)u(t,y)\Big) = \frac{1}{4}\Big(\frac{|y-x|}{2} - t\Big)^2 \mathbf{1}_{\{|y-x|<2t\}}. \tag{7.19}$$

Proof. It following by (7.9), by taking $t = s$. $\qquad\square$

So,

$$\mathbf{E}u(t,x)u(t,y) = f(|x-y|)$$

with

$$f(x) = \frac{1}{16}\mathbf{1}_{(x<2t)}(2t-x)^2 = \frac{1}{16}\begin{cases} 2t-x, x < 2t \\ 0, x \geq 2t. \end{cases}$$

Notice that f is a continuous function on $\mathbb{R}_+$.

We can also express the spatial covariance of the solution (7.4) by using the Fourier transform of the Green kernel (7.3). If $t \geq 0$ and $x, y \in \mathbb{R}$, by (5.10) and (7.3),

$$\mathbf{E}u(t,x)u(t,y) = \int_0^t \int_{\mathbb{R}} G_1(t-u, y-z)G_1(t-u, y-z)dzdu$$

$$= (2\pi)^{-1}\int_0^t du \int_{\mathbb{R}} d\xi e^{-i\langle \xi, x-y\rangle}\frac{\sin((t-u)\|\xi\|)^2}{\|\xi\|^2}.$$

Let us deduce some consequences.

Corollary 7.2. *Let $u(t,x), t \geq 0, x \in \mathbb{R})$ defined by (7.4). Then*

(1) The random field $(u(t,x), x \in \mathbb{R})$ is stationary.
(2) If $|x-y| \geq 2t$, then $u(t,x)$ and $u(t,y)$ are independent random variables.

Proof. We see from (7.19) that the spatial covariance $\mathbf{E}u(t,x)u(t,y)$ is a function of $|x-y|$. Then we apply Proposition 1.2 to get point 1. For point 2., we notice that by (7.19), $u(t,x)$ and $u(t,y)$ are uncorrelated when $|x-y| \geq 2t$, so they are independent. $\qquad\square$

We have a sharp estimates concerning the L^2-norm of the spatial increments.

Lemma 7.2. *Let $u(t,x), t \geq 0, x \in \mathbb{R})$ defined by (7.4). Let T, $M > 0$ and fix $t \in (0,T]$. Then there exist two constants $0 < C_{2,t} < C_1$ such that:*

$$C_{2,t}|x - y| \leq \mathbf{E}\left(|u(t,x) - u(t,y)|^2\right) \leq C_1|x - y| \tag{7.20}$$

for every distinct $x, y \in [-M, M]$.

Proof. We start with the upper bound, so for a fixed $t \in (0,T]$ and for every distinct $x, y \in [-M, M]$, we can compute by the aid of (2.25),

$$\mathbf{E}\left(\,|u(t,x) - u(t,y)|^2\,\right)$$

$$= \int_0^t \int_{\mathbb{R}} \left(G_1(t - s, x - z) - G_1(t - s, y - z)\right)^2 dz\,ds$$

$$= C \int_0^t ds \int_{\mathbb{R}} \left(\frac{\sin((t-s)\|\xi\|)}{\|\xi\|}\right)^2 |e^{-i\xi x} - e^{-i\xi y}|^2 d\xi$$

$$= C \int_0^t ds \int_0^{+\infty} \left(\frac{\sin((t-s)\|\xi\|)}{\xi}\right)^2 |e^{-i\xi(x-y)} - 1|^2 d\xi$$

$$= C \int_0^t ds \int_0^{+\infty} \left(\frac{\sin((t-s)\|\xi\|)}{\xi}\right)^2 \left(2 - 2\cos(\xi(x - y))\right) d\xi$$

$$\leq C \int_0^t ds \int_0^{+\infty} \frac{1 - \cos(\xi(x - y))}{\xi^2} d\xi$$

$$\leq C \int_0^{+\infty} \frac{1 - \cos(\xi(x - y))}{\xi^2} d\xi.$$

Setting $a = \frac{x-y}{|x-y|}$ and making use of the change of variables $p = |x - y|\xi$, the above integral can be expressed:

$$\mathbf{E}\left(\,|u(t,x) - u(t,y)|^2\,\right) \leq C|x - y| \int_0^{+\infty} \frac{1 - \cos(ap)}{p^2} dp.$$

It is easy to see that $\int_0^{+\infty} \frac{1-\cos(ap)}{p^2} dp$ is bounded by a positive constant, indeed applying the fact that $1 - \cos(x) \leq x^2$ and $0 \leq 1 - \cos(x) \leq 2$, we can show that:

$$\int_0^{+\infty} \frac{1 - \cos(ap)}{p^2} dp = \int_0^1 \frac{1 - \cos(ap)}{p^2} dp + \int_1^{+\infty} \frac{1 - \cos(ap)}{p^2} dp$$

$$\leq a^2 + 2 \leq 3.$$

The second part of the proof consists in deducing the lower bound. We split foremost the integral into two regions and we get in this situation:

$$\mathbf{E}\Big(|u(t,x) - u(t,y)|^2\Big)$$

$$= C \int_0^t ds \int_0^1 \left(\frac{\sin((t-s)\|\xi\|)}{\xi}\right)^2 \Big(1 - \cos(\xi(x-y))\Big)d\xi$$

$$+ C \int_0^t ds \int_1^{+\infty} \left(\frac{\sin((t-s)\|\xi\|)}{\xi}\right)^2 \Big(1 - \cos(\xi(x-y))\Big)d\xi.$$

As the first quantity is a positive one, we can neglect its contribution, then by Fubini's theorem:

$$\mathbf{E}\Big(\,|u(t,x) - u(t,y)\,|^2\,\Big)$$

$$\geq C \int_0^t ds \int_1^{+\infty} \left(\frac{\sin((t-s)\|\xi\|)}{\xi}\right)^2 \Big(1 - \cos(\xi(x-y))\Big)d\xi$$

$$\geq C \int_1^{+\infty} \frac{1 - \cos(\xi(x-y))}{\xi^2} \left(t - \frac{1}{2\|\xi\|}\sin(2t|\xi|)\right) d\xi.$$

Without loss of generality we may assume now that $t = 1$, and as $1 - \frac{\sin(2|\xi|)}{2\|\xi\|} \geq \frac{1}{2}$, for $\xi \geq 1$ and $1 - \cos(x) = 2\sin^2(\frac{x}{2})$ for every real x, it yields that:

$$\mathbf{E}\Big(\,|u(t,x) - u(t,y)\,|^2\,\Big) \geq C \int_1^{+\infty} \frac{1 - \cos(\xi(x-y))}{\xi^2}d\xi$$

$$\geq C \int_1^{+\infty} \frac{\sin^2(a\xi\frac{|x-y|}{2})}{\xi^2}d\xi$$

$$\geq C\,|x-y|\int_{|x-y|/2}^{+\infty} \frac{\sin^2(aq)}{q^2}dq$$

$$\geq C\,|x-y|\int_M^{+\infty} \frac{\sin^2(aq)}{q^2}dq$$

$$= C_{2,1}\,|x-y|\,.$$

In the second and the third inequalities above, we again denoted by $a = \frac{x-y}{|x-y|}$ and we made the change of variables by $q = \frac{\xi|x-y|}{2}$, and at the last line we used the fact that the integral $\int_M^{+\infty} \frac{\sin^2(u)}{u^2}du$ is finite and independent of x and y, which establishes (7.20). $\qquad\square$

Remark 7.5.

(1) Lemma 7.2 implies the (sharp) $\frac{1}{2}$-Hölder regularity in space of the solution. We notice that this coincides with the Hölder regularity order in time, that is, the solution to the wave equation has the same regularity in time and in space. This is not the case for the stochastic heat equation.

(2) It is also possible to obtain a (quasi) similar double inequality for small increments by using directly the expression of the Green kernel (7.2) instead of its Fourier transform. By (7.19), for every $x, y \in \mathbb{R}$ and $t \geq 0$, such that $|x - y| < 2t$,

$$
\begin{aligned}
\mathbf{E}\left(u(t,x) - u(t,y)\right)^2 &= \mathbf{E}u(t,x)^2 + \mathbf{E}u(t,y)^2 - 2\mathbf{E}u(t,x)u(t,y) \\
&= \frac{1}{8}\left(4t^2 - (2t - |x-y|)^2\right) \\
&= \frac{1}{8}|x-y|(4t - |x-y|).
\end{aligned}
\tag{7.21}
$$

Then,

$$
\begin{aligned}
\mathbf{E}\left(u(t,x) - u(t,y)\right)^2 &\leq \frac{1}{8}|x-y|(4t + |x-y|) \\
&\leq \frac{1}{8}6t|x-y| = \frac{3t}{4}|x-y|.
\end{aligned}
$$

The above bound precises the constant C_1 in (7.20) and it does not require to have x, t in an interval of type $[-M, M]$. On the other hand, we assumed $|x - y| < 2t$.

Also, if we further assume $|x - y| < t$, so $4t - |x - y| > 3t$, we will have

$$
\mathbf{E}\left(u(t,x) - u(t,y)\right)^2 \geq \frac{3t}{8}|x-y|,
$$

and this gives the concrete form for the constant $C_{2,t}$ in (7.20).

Let us discuss the link between $(u(t,x), x \in \mathbb{R})$ and the Wiener process. This link will be seen through the increment process generated by the solution u. For $t \geq 0$ and $x \in \mathbb{R}$, define

$$
X(t,x) = u(t, x+1) - u(t,x).
$$

This is the spatial "noise" generated in space by the solution u (recall also Definition 2.2 for the noise generated by the fBm).

Recall that the two-sided Wiener process $(W(x), x \in \mathbb{R})$ is a Gaussian process with covariance (2.9).

Proposition 7.7. *Let* u *be given by (7.4) and assume* $t > 1$. *Let* $(W(x), x \in \mathbb{R})$ *be a two-sided Brownian motion and let* $Z \sim N(0,1)$ *independent by* u. *Then*

$$\left(X(t,x) + \frac{1}{2\sqrt{2}} Z, x \in [0,1] \right) \equiv^{(d)} \left(\sqrt{\frac{t}{2}} (W(x+1) - W(x)), x \in [0,1] \right).$$
$$(7.22)$$

Proof. Assume $x, y \in [0,1]$ and $t > \frac{1}{2}$. Then $|x - y| < 2t$, $|x - y + 1| < 2t$, $|x - y - 1| < 2t$ and by (7.19),

$$\mathbf{E} X(t,x) X(t,y)$$
$$= \mathbf{E}(u(t,x+1) - u(t,x))(u(t,y+1) - u(t,y))$$
$$= \frac{1}{16} \left[2(2t - |x-y|)^2 - (2t - |x-y+1|)^2 - (2t - |x-y-1|)^2 \right]$$
$$= \frac{1}{16} 4t \left[-2|x-y| + |x-y+1| + |x-y-1| \right]$$
$$+ \frac{1}{16} \left[2|x-y|^2 - |x-y+1|^2 - |x-y-1|^2 \right]$$

so, since $2|x-y|^2 - |x-y+1|^2 - |x-y-1|^2 = -2$

$$\mathbf{E} X(t,x) X(t,y) = \frac{t}{4} \left[-2|x-y| + |x-y+1| + |x-y-1| \right] - \frac{1}{8}. \quad (7.23)$$

Denote, for $t > 0, x \in [0,1]$,

$$Y(t,x) := X(t,x) + \frac{1}{2\sqrt{2}} Z.$$

Then by (7.23),

$$\mathbf{E} Y(t,x) Y(t,y) = \frac{t}{4} \left[-2|x-y| + |x-y+1| + |x-y-1| \right]. \quad (7.24)$$

Let $(W(x), x \in \mathbb{R})$ be a two-sided Wiener process. From (2.9), for every $x, y \in \mathbb{R}$,

$$\mathbf{E}(W(x+1) - W(x))(W(y+1) - W(y))$$
$$= \frac{1}{2} (|x - y + 1| + |x - y - 1| - 2|x - y|). \quad (7.25)$$

By (7.24) and (7.25) we obtain the conclusion (7.22). $\qquad \square$

We have the following estimate for the joint increment of the solution.

Proposition 7.8. *For every* $s, t \in [0, T]$ *and for every* $x, y \in [-M, M]$ *with* $M > 0$, *we have*

$$\mathbf{E}|u(t,x) - u(s,y)|^2 \leq C \left(|t - s| + |x - y| \right). \quad (7.26)$$

Therefore, the solution u *admits a bicontinuous version.*

Proof. The bound (7.26) follows from point 3. in Proposition 7.4 and by the upper bound in Lemma 7.2. Then, (7.26) implies for every $p \geq 2$,

$$\mathbf{E}|u(t,x) - u(s,y)|^p \leq C \left(|t-s| + |x-y|\right)^{\frac{p}{2}}$$
$$\leq C \left(|t-s|^{\frac{p}{2}} + |x-y|^{\frac{p}{2}}\right).$$

The existence of a bicontinuous version in (t,x) follows from Theorem 3.2.

$\square$

PART 3

Power variation and statistical inference for solutions to SPDEs

Chapter 8

Variations of the solution to the stochastic heat equation

We will use the distributional and trajectorial properties of the solution to the heat and wave equation in order to estimate the parameter that may appear in these equations. The estimators defined in the next chapters will be constructed by using the p-variations of the solution. We start by analyzing the limit behavior of these p-variations.

8.1 Exact and renormalized variations of the perturbed fractional Brownian motion

We first define the concept of *exact q-variation* for stochastic processes. This notion has been already introduced for fractional and bifractional Brownian motion in (2.15) and (2.29).

Definition 8.1. Let $A_1 < A_2$ and for $n \geq 0$, let $t_i = A_1 + \frac{i}{n}(A_2 - A_1)$ for $i = 0, \ldots, n$. A continuous stochastic process $(X_t, t \geq 0)$ admits an exact q-variation (or a variation of order q) over the interval $[A_1, A_2]$ if the sequence

$$S_{[A_1, A_2]}^{n,q}(X) = \sum_{i=0}^{n-1} \left| X_{t_{i+1}} - X_{t_i} \right|^q \tag{8.1}$$

converges in probability as $n \to \infty$. The limit, when it exists, is called the exact q-variation of X over the interval $[A_1, A_2]$.

If $[A_1, A_2] = [0, t]$, we will simply denote $S_{[0,t]}^{n,q}(X) := S_t^{n,q}(X)$. Moreover, if $t = 1$, we denote $S_t^{N,q}(X) := S^{q,N}$. In the case $q = 2$ the limit of $S^{2,n}$ is called the quadratic variation, for $q = 3$ we have the cubic variation while for $q = 4$ the quartic variation.

For fractional and bifractional Brownian motion, we discussed the behavior of their q-variations in Chapter 2, see Theorem 2.1 and Theorem

2.3. Let us recall the result (2.30) for the bi-fBm (which includes, as a particular case, the fBm).

Proposition 8.1. *If $(B_t^{H,K}, t \geq 0)$ is a bi-fBm with Hurst parameters $H \in (0,1), K \in (0,1]$ then $B^{H,K}$ admits a variation of order $\frac{1}{HK}$ over any interval $[A_1, A_2]$ which is equal to $C_{H,K}(A_2 - A_1)$, i.e.*

$$S_{[A_1,A_2]}^{N,\frac{1}{HK}}(B^{H,K}) \to_{N \to \infty} C_{H,K}(A_2 - A_1)$$

where

$$C_{H,K} = 2^{\frac{1-K}{2}} \mathbf{E}|Z|^{\frac{1}{HK}} \tag{8.2}$$

with $Z \sim N(0,1)$ a standard normal random variable.

By taking $K = 1$, we notice that the fractional Brownian motion has $\frac{1}{H}$-variation over the interval $[A_1, A_2]$ given by $\mathbf{E}|Z|^{\frac{1}{H}}(A_2 - A_1)$.

The purpose is to obtain the exact q-variation for solutions to stochastic heat and wave equations. First, we deal with the situation of the stochastic heat equation. We recall that the solution to the (standard or fractional) heat equation can be decomposed as the sum of a fBm (with Hurst parameter less or equal than one half) and a smooth process. This is both available for the spatial or temporal process. Therefore, we need to understand the variations of such stochastic processes, i.e. the sum of a fBm with Hurst index H and of another smooth Gaussian process which satisfies (8.3) or it has absolutely continuous paths (this will be called in the sequel as *perturbed fractional Brownian motion* with Hurst index H).

Let us state the result concerning the asymptotic behavior of the exact q-variation of the sum of the perturbed fBm.

Lemma 8.1. *Let $(B_t^H, t \geq 0)$ be a fBm with $H \in (0,1)$ and consider a centered Gaussian process $(X_t, t \geq 0)$ such that*

$$\mathbf{E}|X_t - X_s|^2 \leq C|t - s|^2 \text{ for every } s, t \geq 0. \tag{8.3}$$

Define

$$Y_t^H = B_t^H + X_t \text{ for every } t \geq 0. \tag{8.4}$$

Then the process Y^H has $\frac{1}{H}$-variation over the interval $[A_1, A_2]$ which is equal to

$$\mathbf{E}|Z|^{1/H}(A_2 - A_1),$$

i.e.

$$S_{[A_1,A_2]}^{n,\frac{1}{H}}(Y^H) \to_{n \to \infty} \mathbf{E}|Z|^{\frac{1}{H}}(A_2 - A_1) \text{ in probability,}$$

where $S_{[A_1,A_2]}^{n,\frac{1}{H}}(Y^H)$ is given by (8.1).

Proof. We use Minkowski's inequality (2.31) to write

$$\left(\sum_{i=0}^{n-1}\left|B_{t_{i+1}}^H - B_{t_i}^H\right|^{\frac{1}{H}}\right)^H - \left(\sum_{i=0}^{n-1}\left|X_{t_{i+1}} - X_{t_i}\right|^{\frac{1}{H}}\right)^H$$

$$\leq \left(\sum_{i=0}^{n-1}\left|Y_{t_{i+1}}^H - Y_{t_i}^H\right|^{\frac{1}{H}}\right)^H$$

$$\leq \left(\sum_{i=0}^{n-1}\left|B_{t_{i+1}}^H - B_{t_i}^H\right|^{\frac{1}{H}}\right)^H + \left(\sum_{i=0}^{n-1}\left|X_{t_{i+1}} - X_{t_i}\right|^{\frac{1}{H}}\right)^H \quad (8.5)$$

Since by Theorem 2.1, the sequence

$$\sum_{i=0}^{n-1}\left|B_{t_{i+1}}^H - B_{t_i}^H\right|^{\frac{1}{H}}$$

converges in probability, as $n \to \infty$, to the desired limit $\mathbf{E}|Z|^{1/H}(A_2 - A_1)$, it suffices to show that

$$\sum_{i=0}^{n-1}\left|X_{t_{i+1}} - X_{t_i}\right|^{\frac{1}{H}} \text{ converges in probability to zero.}$$

We have, via (8.3) and the fact that for $s < t$,

$$X_t - X_s \sim N(0, \sigma(s,t)^2) \text{ with } \sigma(s,t)^2 \leq C(t-s)^2$$

$$\mathbf{E}\sum_{i=0}^{n-1}\left|X_{t_{i+1}} - X_{t_i}\right|^{\frac{1}{H}} = C\sum_{i=0}^{n-1}\left(\mathbf{E}\left|X_{t_{i+1}} - X_{t_i}\right|^2\right)^{\frac{1}{2H}}$$

$$\leq C\sum_{i=0}^{n-1}\left|t_{i+1} - t_i\right|^{\frac{1}{H}} \leq Cn^{1-\frac{1}{H}} \to_{n\to\infty} 0.$$

$\square$

We have a similar result if we replace condition (8.3) by the absolute continuity of the paths of X.

Lemma 8.2. *Let $(B_t^H, t \geq 0)$ be a fBm with $H \in (0,1)$ and consider a centered Gaussian process $(X_t, t \geq 0)$ with absolute continuous sample paths. Define $(Y_t^H, t \geq 0)$ by (8.4). Then the process Y^H has $\frac{1}{H}$-variation over the interval $[A_1, A_2]$ which is equal to*

$$\mathbf{E}|Z|^{1/H}(A_2 - A_1).$$

Proof. We have the double bound (8.5). Now

$$\sum_{i=0}^{n-1} \left|X_{t_{i+1}} - X_{t_i}\right|^{\frac{1}{H}} \leq \sup_{i=0,\dots,N-1} \left(\left|X_{t_{i+1}} - X_{t_i}\right|^{\frac{1}{H}-1}\right) \sum_{i=0}^{n-1} \left|X_{t_{i+1}} - X_{t_i}\right|.$$

The quantity $\sup_{i=0,1,\dots,n-1} \left|X_{t_{i+1}} - X_{t_i}\right|^{\frac{1}{H}-1}$ converges to zero almost surely as $n \to \infty$ due to the absolute continuity of X while the sum is bounded by the total variation of X on the interval $[A_1, A_2]$. $\qquad \square$

Remark 8.1. Lemmas 8.1 and 8.2 show that the perturbed fBm Y^H given by (8.4) has the same q-variation as the fBm B^H. The regular perturbation X (which is absolute continuous or it verifies (8.3)) does not affect the behavior of these variations.

We will also analyze the asymptotic behavior of the renormalized q-variation for more general stochastic processes, not necessarily self-similar. For a general process $(X_t, t \geq 0)$, by studying the asymptotic behavior of *the renormalized q-variation* we will generally mean to find a constant $\mu \in \mathbb{R}$ and two deterministic sequence $f(n), g(n)$ which converge to zero as $n \to \infty$ such that $\mathbf{E}\left[f(n)^{-1}\left(X_{t_{i+1}} - X_{t_i}\right)\right]^2$ is close to 1 and (we use the notation $\mu_q = \mathbf{E}(Z^q)$ with $Z \sim N(0,1)$)

$$g(n) \sum_{i=0}^{n-1} \left[\frac{\left(X_{t_{i+1}} - X_{t_i}\right)^q}{f(n)^q} - \mu_q\right]$$

converges in distribution to a non-trivial limit as $n \to \infty$. If X is a H-self-similar process (for example, the fractional Brownian motion) and $[A_1, A_2] = [0, 1]$, then the natural choice for f is $f(n) = n^{-H}$. But we can define the concept of renormalized q-variation for the perturbed fBm.

We recall the following result concerning the variations of the fractional Brownian motion (see [Breuer and Major (1993)], [Dobrushin and Major (2005)], [Giraitis and Surgailis (1990)], [Nourdin and Peccati (2012)], [Taqqu (1975)] or [Tudor (2013)]; the reader may also consult Section 1 in [Nourdin et al. (2010)] for a survey of these results). We will restrict below to the case when the Hurst parameter H is less or equal than $\frac{1}{2}$ since only this case will be needed in the sequel (but similar results are available for H in the whole interval $(0, 1)$).

Theorem 8.1. *Let* $(B_t^H, t \geq 0)$ *be a fBm with* $H \in (0, \frac{1}{2})$ *and consider an integer number* $q \geq 2$. *Let* $t_i, i = 0, 1, \dots, n$ *be given by (2.14). Define, for* *every $n \geq 1$,*

$$V_{q,n,[A_1,A_2]}(B^H) := \sum_{i=0}^{n-1} \left[\frac{n^{Hq}}{(A_2 - A_1)^{Hq}} (B_{t_{i+1}}^H - B_{t_i}^H)^q - \mu_q \right].$$

Then

$$\frac{1}{\sqrt{n}} V_{q,n,[A_1,A_2]}(B^H) \to^{(d)} N(0, \sigma_{H,q}^2). \tag{8.6}$$

The exact expression of the limit variance $\sigma_{H,q}^2$ can be found in e.g. [Nourdin and Peccati (2012)]. We will apply Theorem 8.1 to obtain the limit behavior in distribution of the renormalized q-variation of the perturbed fBm. For a perturbed fBm $(U_t, t \geq 0)$ with Hurst parameter $H \in (0,1)$, we set

$$V_{q,n,[A_1,A_2]}(U) := \sum_{i=0}^{n-1} \left[\frac{n^{Hq}}{(A_2 - A_1)^{Hq}} (U_{t_{i+1}} - U_{t_i})^q - \mu_q \right]. \tag{8.7}$$

where $(t_i, i = 0, 1, \ldots, n)$ is the partition (2.14) of the interval $[A_1, A_2]$.

Lemma 8.3. *Let $(B_t^H, t \geq 0)$ be a fBm with $H \in \left(0, \frac{1}{2}\right)$ and consider a centered Gaussian process $(X_t, t \geq 0)$ satisfying (8.3). Define $(Y_t^H, t \geq 0)$ by (8.4) and let*

$$V_{q,n,[A_1,A_2]}(Y^H) := \sum_{i=0}^{n-1} \left[\frac{n^{Hq}}{(A_2 - A_1)^{qH}} (Y_{t_{i+1}}^H - Y_{t_i}^H)^q - \mu_q \right]. \tag{8.8}$$

Then if $H \in (0, \frac{1}{2})$ and $q \geq 2$ is an integer, as $n \to \infty$,

$$\frac{1}{\sqrt{n}} V_{q,n,[A_1,A_2]}(Y^H) \tag{8.9}$$

$$= \frac{1}{\sqrt{n}} \sum_{i=0}^{n-1} \left[\frac{n^{Hq}}{(A_2 - A_1)^{qH}} (Y_{t_{i+1}}^H - Y_{t_i}^H)^q - \mu_q \right] \to^{(d)} N(0, \sigma_{H,q}^2)$$

with $\sigma_{H,q}^2$ from (8.6).

Proof. By using Newton's formula we can write

$$V_{q,n,[A_1,A_2]}(Y^H) = V_{q,n,[A_1,A_2]}(B^H) + R_n$$

where

$$R_n = \frac{n^{Hq}}{\sqrt{n}} \sum_{r=0}^{q-1} C_q^r \sum_{i=0}^{n-1} (B_{t_{i+1}}^H - B_{t_i}^H)^r (X_{t_{i+1}} - X_{t_i})^{q-r} := \sum_{r=0}^{q-1} R_{n,r}. \tag{8.10}$$

It suffices to show that $\frac{1}{\sqrt{n}}R_{n,r}$ converges to zero in $L^1(\Omega)$ for every $r = 0, \ldots, q-1$. Using (8.3) we have for every $s, t \geq 0$ and for $r = 0, \ldots, q-1$

$$\mathbf{E}|X_t - X_s|^{2(q-r)} \leq C|t - s|^{2(q-r)}$$

and then we can write, for $r = 0, \ldots, q-1$, via Cauchy-Schwarz's inequality,

$$\mathbf{E}\frac{1}{\sqrt{n}}|R_{n,r}| \leq \frac{1}{\sqrt{n}} \sum_{i=0}^{n-1} \left(\mathbf{E}(B_{t_{i+1}}^H - B_{t_i}^H)^{2r}\right)^{\frac{1}{2}} \left(\mathbf{E}(X_{t_{i+1}} - X_{t_i})^{2(q-r)}\right)^{\frac{1}{2}}$$

$$\leq Cn^{(H-1)(q-r)+\frac{1}{2}} \leq Cn^{H-\frac{1}{2}}$$

and this converges to zero as $n \to \infty$ since $H < \frac{1}{2}$. □

When $H = \frac{1}{2}$, then $Y^H = Y^{\frac{1}{2}}$ is a perturbed Brownian motion. We can obtain the asymptotic behavior of its (exact and renormalized) quadratic variation under an additional hypothesis on the perturbation X.

Lemma 8.4. *Let $(B_t, t \geq 0)$ be a Wiener process and assume that X is a centered Gaussian process which verifies (8.3). Assume in addition that X is independent of B and let $(Y_t^H, t \geq 0)$ be given by (8.4). Then*

$$\frac{1}{\sqrt{n}}V_{2,n,[A_1,A_2]}(Y^{\frac{1}{2}}) = \frac{1}{\sqrt{n}} \sum_{i=0}^{n-1} \left[\frac{n^{2H}}{(A_2 - A_1)^{2H}}(Y_{t_{i+1}}^H - Y_{t_i}^H)^2 - 1\right]$$

$$\to^{(d)} N(0, \sigma_{\frac{1}{2},2}^2). \tag{8.11}$$

Proof. If $H = \frac{1}{2}$ and $q = 2$, we have

$$\frac{1}{\sqrt{n}}V_{2,n,[A_1,A_2]}(Y^{\frac{1}{2}})$$

$$= \frac{1}{\sqrt{n}}V_{2,n,[A_1,A_2]}(B) + \frac{2}{\sqrt{n}} \sum_{i=0}^{n-1} \frac{n}{A_2 - A_1}(B_{t_{i+1}} - B_{t_i})(X_{t_{i+1}} - X_{t_i})$$

$$+ \frac{1}{\sqrt{n}} \frac{n}{A_2 - A_1} \sum_{i=0}^{n-1} (X_{t_{i+1}} - X_{t_i})^2.$$

Clearly, by (8.3)

$$\sqrt{n}\mathbf{E} \sum_{i=0}^{n-1} (X_{t_{i+1}} - X_{t_i})^2 \leq cn^{-\frac{1}{2}} \to_{n\to\infty} 0$$

and since X is independent by B, we have for $i \neq j$,

$$\mathbf{E}\left[(B_{t_{i+1}} - B_{t_i})(X_{t_{i+1}} - X_{t_i})(B_{t_{j+1}} - B_{t_j})(X_{t_{i+1}} - X_{t_i})\right] = 0. \tag{8.12}$$

Thus, by (8.12) and (8.3),

$$\mathbf{E}\left(\sqrt{n}\sum_{i=0}^{n-1}(B_{t_{i+1}}-B_{t_i})(X_{t_{i+1}}-X_{t_i})\right)^2$$

$$= n\mathbf{E}\sum_{i=0}^{n-1}(B_{t_{i+1}}-B_{t_i})^2(X_{t_{i+1}}-X_{t_i})^2$$

$$= n\sum_{i=0}^{n-1}\mathbf{E}(B_{t_{i+1}}-B_{t_i})^2 E(X_{t_{i+1}}-X_{t_i})^2$$

$$= n\sum_{i=0}^{n-1}\mathbf{E}(X_{t_{i+1}}-X_{t_i})^2$$

$$\leq C\frac{1}{n}\to_{n\to\infty} .$$

$\square$

Lemma 8.4 indicates that the perturbation X which satisfies (8.3) does not affect the renormalized q-variation of the fBm. This result still holds if we suppose that the process X has absolutely continuous trajectories.

Lemma 8.5. *Let $(B_t^H, t \geq 0)$ be a fBm with $H \in (0, \frac{1}{2})$ and consider a centered Gaussian process $(X_t, t \geq 0)$ with absolutely continuous sample paths. Let $(Y_t^H, t \geq 0)$ be the perturbed fBm (8.4) and let $V_{q,n,[A_1,A_2]}(Y^H)$ be given by (8.8). Then if $H \in (0, \frac{1}{2})$ and $q \geq 2$ is an integer*

$$\frac{1}{\sqrt{n}}V_{q,n,[A_1,A_2]}(Y^H) \to^{(d)} N(0, \sigma_{H,q}^2). \tag{8.13}$$

Proof. With $R_{n,r}$ $(r \geq 0)$ given by (8.10)

$$\mathbf{E}\frac{1}{\sqrt{n}}|R_{n,r}|$$

$$\leq \left(\sup_{i=0,1,\dots,n-1}|X_{t_{i+1}}-X_{t_i}|^{q-r-1}\left|B_{t_{i+1}}^H-B_{t_i}^H\right|^r\right)\sum_{i=0}^{n-1}|X_{t_{i+1}}-X_{t_i}|$$

and by the continuity of X and B^H for $r = 0, 1, \dots, q-1$ the factor $\sup_i|X_{t_{i+1}}-X_{t_i}|^{q-r-1}\left|B_{t_{i+1}}^H-B_{t_i}^H\right|^r$ converges to zero almost surely as $n \to \infty$. On the other hand, $\sum_{i=0}^{n-1}|X_{t_{i+1}}-X_{t_i}|$ is bounded by the total variation of X over the interval $[A_1, A_2]$. The conclusion is obtained as in Lemma 10.20. $\square$

As a corollary, we give the exact and renormalized variation of the bifractional Brownian motion. We insist on the following point: the notion of renormalized q-variation has been defined in (8.7) only for perturbed fractional Brownian motion (not a general stochastic process). It particular, it applies to bifractional Brownian motion as, we will se later, to the solution to the stochastic heat equation. As usual, the notation " $\to^{(d)}$ " refers to the convergence in distribution.

Corollary 8.1. *Let* $(B_t^{H,K}, t \geq 0)$ *be a bifractional Brownian motion with* $HK \in \left(0, \frac{1}{2}\right)$. *Then*

(1) With $S_{[A_1,A_2]}^{N,q}(B^{H,K})$ *from Definition 8.1,*

$$S_{[A_1,A_2]}^{N,\frac{1}{HK}}(B^{H,K}) \to_{N\to\infty} (A_2 - A_1)2^{\frac{1-K}{2}}\mathbf{E}|Z|^{\frac{1}{HK}} \text{ in probability.}$$

(2) Let $V_{q,N,[A_1,A_2]}$ *be given by (8.7). Then*

$$\frac{1}{\sqrt{n}}V_{\frac{1}{HK},N,[A_1,A_2]}((C_2^{-1}B^{H,K}) \to_{N\to\infty}^{(d)} N(0,\sigma_{H,q}^2)$$

with $\sigma_{H,q}$ *from (8.6) and* $C_2 = 2^{\frac{1-K}{2}}$.

Proof. Point 1. is exactly Proposition 8.1. It also follows from the decomposition of the bi-fBm in Theorem 2.2 using the fact that the process $X^{H,K}$ is absolutely continuous, by applying Lemma 8.2. Point 2. is also a consequence of the decomposition in Theorem 2.2 and of Lemma 8.5 (see Proposition 8.3), because $C_2^{-1}B^{H,K}$ is a perturbed fractional Brownian motion with Hurst index $HK \in \left(0, \frac{1}{2}\right)$. $\qquad\square$

8.2 Variations of the isotropic fractional Brownian motion

Let $(B^H(x), x \in \mathbb{R}^d)$ be an isotropic fractional Brownian sheet with Hurst index $H \in (0,1)$. The basic properties of this Gaussian process have been presented in Section 3.3.

Following the one-parameter case, we define the q-variation of the isotropic fBm as the limit in probability as $n \to \infty$, of the sequence

$$S_{[A_1,A_2]}^{n,q}(B^H) = \sum_{i=0}^{n-1} |B^H(\mathbf{x}_{i+1}) - B^H(\mathbf{x}_i)|^q$$

where $A_1, A_2 \in \mathbb{R}$ with $A_1 < A_2$, $\mathbf{x}_i = (x_i^{(1)}, \ldots, x_i^{(d)})$ with $x_i^{(j)} = A_1 + \frac{i}{n}(A_2 - A_1)$ for $i = 0, \ldots, n$ and $j = 1, \ldots, d$.

Let us state the result on the limit behavior of the q-exact variation of the isotropic fractional Brownian sheet. Its proof follows easily from the one-parameter case.

Proposition 8.2. *The isotropic fBm* $(B^H(\mathbf{x}), \mathbf{x} \in \mathbb{R}^d)$ *has* $\frac{1}{H}$*-variation over* $[A_1, A_2]$ *which is equal to*

$$(A_2 - A_1)\mathbf{E}|B_1|^{1/H} = (A_2 - A_1)d\mathbf{E}|Z|^{1/H}$$

Proof. Consider the sequence

$$Y_{n,q} = n^{qH-1} \sum_{i=0}^{n-1} \left|B^H(\mathbf{x}_{i+1}) - B^H(\mathbf{x}_i)\right|^q.$$

From Proposition Proposition 3.6 and Proposition 3.7, it has the same law as

$$Y'_{n,q} = (A_2 - A_1)^{qH} \frac{1}{n} \sum_{i=0}^{n-1} \left|B^H(\mathbf{j}+1) - B^H(\mathbf{j})\right|^q.$$

with $\mathbf{j} = (j, \ldots, j) \in \mathbb{R}^d$. The sequence

$$\left(B^H(\mathbf{j}+1) - B^H(\mathbf{j}), \mathbf{j} \in \mathbb{Z}^d\right)$$

is stationary and it has the same law as $d^H(B_{j+1} - B_j)$ where B is a one-parameter fBm with Hurst parameter H.

By the ergodic theorem (see e.g. Chapter 2 in [Samorodnitsky (2016)], see also (2.16)), $Y'_{n,q}$ converges an $n \to \infty$ to $(A_2 - A_1)^{qH}\mathbf{E}|B_1|^q$ almost surely and in $L^1(\Omega)$. Taking $q = \frac{1}{H}$, we obtain the conclusion. □

Following the proof of Lemma 8.1, we can obtain the q-variation of the isotropic fBm perturbed by a regular multiparameter process.

Lemma 8.6. *Let* $(B^H(\mathbf{x}), \mathbf{x} \in \mathbb{R}^d)$ *be a d-parameter isotropic fBm with* $H \in \left(\frac{1}{2}, 1\right)$ *and consider* $(X(x), x \in \mathbb{R}^d)$ *a d-parameter stochastic process that satisfies*

$$\mathbf{E}\left(X(\mathbf{x}) - X(\mathbf{y})\right|^2 \le C\|\mathbf{x} - \mathbf{y}\|^2, \text{ for every } \mathbf{x}, \mathbf{y} \in \mathbb{R}^d. \tag{8.14}$$

Define

$$Y(\mathbf{x}) = B^H(\mathbf{x}) + X(\mathbf{x}) \text{ for every } \mathbf{x} \in \mathbb{R}^d.$$

Then

(1) the process $(Y(\mathbf{x}), \mathbf{x} \in \mathbb{R}^d)$ has $\frac{1}{H}-$ exact variation on $[A_1, A_2]$ which is equal to

$$(A_2 - A_1) d\mathbf{E}|Z|^{1/H}.$$

(2) Then if $H \in (0, \frac{1}{2})$ and $q \geq 2$ is an integer, with $\sigma^2_{H,q}$ from (8.6),

$$\frac{1}{\sqrt{n}} V_{q,n}(Y^H) \tag{8.15}$$

$$= \frac{1}{\sqrt{n}} \sum_{i=0}^{n-1} \left[\frac{n^{Hq} d^{-Hq}}{(A_2 - A_1)^{qH}} (Y^H(\mathbf{x}_{i+1}) - Y^H(\mathbf{x}_i))^q - \mu_q \right]$$

$$\rightarrow_{n \to \infty} N(0, \sigma^2_{H,q}).$$

Proof. As in the proof of Lemma 8.1, we have the double inequality (8.5) due to Minkovski inequality. The sequence

$$\sum_{i=0}^{n-1} |B^H(\mathbf{x}_{i+1}) - B^H(\mathbf{x}_i)|^{\frac{1}{H}}$$

converges again almost surely and in L^1 to the desired limit, by Proposition 8.2. It remains to show that $\sum_{i=0}^{n-1} |X(\mathbf{x}_{i+1}) - X(\mathbf{x}_i)|^{\frac{1}{H}}$ converges to zero in probability and this is an easy consequence of the hypothesis (8.14).

For point 2. it suffices to observe that the vector

$$\left(B^H(\mathbf{x}_{i+1}) - B^H(\mathbf{x}_i), i = 0, 1, \ldots, n-1 \right)$$

has the same law as

$$\left(d^H(B(x_{i+1}) - B(x_i)), i = 0, 1, \ldots, n-1 \right)$$

where B is a one-parameter fBm with Hurst index H and to apply Lemma 8.1. $\qquad \square$

8.3   Variations of the solution to the heat equation with space-time white noise

For a two-parameter stochastic process $(u(t, x), t \geq 0, x \in \mathbb{R})$, we can consider its variations in time (or its temporal variations) and its variations in space (or its spatial variations). That is, if the space variable $x \in \mathbb{R}$ is fixed and $t_i, i = 0, \ldots, N$ are given by (2.14), then the (exact) temporal q-variation sequence of u is

$$S^{N,q}_{[A_1, A_2]}(u(\cdot, x)) = \sum_{i=0}^{N-1} |u(t_{i+1}, x) - u(t_i, x)|^q. \tag{8.16}$$

Similarly, if $t \geq 0$ is fixed, the spatial variation sequence of u is given by

$$S^{N,q}_{[A_1,A_2]}(u(t,\cdot)) = \sum_{i=0}^{N-1} |u(t,x_{i+1}) - u(t,x_i)|^q \qquad (8.17)$$

where $(x_i, i = 0, 1, \ldots, N)$ forms a partition of $[A_1, A_2]$ with $A_1 < A_2$ (they are not necessarily positive) given by

$$x_i = A_1 + \frac{i}{N}(A_2 - A_1). \qquad (8.18)$$

We consider $(u(t,x), t \geq 0, x \in \mathbb{R})$, the solution to the stochastic (fractional) heat equation, and we analyze separately its temporal and spatial variations.

8.3.1 *Temporal variations*

Let $(u(t,x), t \geq 0, x \in \mathbb{R})$ be the mild solution to the (fractional) heat equation (5.38). We assume that the parameter of the fractional Laplacian satisfies $\alpha \in (1, 2]$. In this way, we include both cases of the fractional heat equation $\alpha \in (1, 2)$ and of the standard heat equation (which corresponds to $\alpha = 2$).

From the result concerning the asymptotic behavior of the p-variations of the bifractional Brownian motion and the link between the law of the solution and the bi-fBm (Proposition 5.5 and Proposition 5.16) we will deduce the behavior of the temporal variations of the mild solution.

We will need the following trivial lemma.

Proposition 8.3. *Let $(X_n, n \geq 0), (Y_n, n \geq 0)$ be two sequences of random variables such that $X_n =^{(d)} Y_n$ for every $n \geq 0$. Assume that $X_n \to_{n\to\infty} L \in \mathbb{R}$ in probability. Then $Y_n \to_{n\to\infty} L$ in probability.*

Proof. Clearly Y_n converges in law to L and $n \to \infty$ and this implies the convergence in probability, since the limit is a constant. □

We will use the fact that two stochastic processes with the same finite dimensional distributions have the same variations. Recall that " $=^{(d)}$ " denotes the equivalence of finite dimensional distributions.

Corollary 8.2. *Let $(X_t), t \geq 0)$ and $(Y_t, t \geq 0)$ be two stochastic process such that*

$$(X_t, t \geq 0) \equiv^{(d)} (Y_t, t \geq 0). \qquad (8.19)$$

Let $(t_i, i = 0, \ldots, N)$ be the partition of the interval $[A_1, A_2]$ given by (2.14). Assume that for $q > 0$

$$S^{N,q}_{[A_1, A_2]}(X) \to_{N \to \infty} L \in \mathbb{R} \text{ in probability.}$$

Then

$$S^{N,q}_{[A_1, A_2]}(Y) \to_{N \to \infty} L \in \mathbb{R} \text{ in probability.}$$

Proof. By (8.19),

$$S^{N,q}_{[A_1, A_2]}(X) =^{(d)} S^{N,q}_{[A_1, A_2]}(Y).$$

Then we apply Proposition 8.3. □

Let us state the result concerning the exact variation in time for the mild solution (5.46).

Proposition 8.4. *Let $(u(t, x), t \geq 0)$ be given by (5.46). Fix $0 \leq A_1 < A_2$ and $x \in \mathbb{R}$. Let $t_j = A_1 + \frac{j}{n}(A_2 - A_1), n \geq 1, j = 0, 1, \ldots, n$ be a partition of the interval $[A_1, A_2]$. Then the process $(u(t, x), t \geq 0)$ admits variation of order $\frac{2\alpha}{\alpha - 1}$ which is equal to*

$$c_{2,\alpha}^{\frac{2\alpha}{\alpha - 1}} C_{\frac{1}{2}, 1 - \frac{1}{\alpha}}(A_2 - A_1)$$

with $C_{\frac{1}{2}, 1 - \frac{1}{\alpha}}(A_2 - A_1)$ from (8.2) and $c_{2,\alpha}$ from (5.49). That is, for every $x \in \mathbb{R}$,

$$S^{N, \frac{2\alpha}{\alpha - 1}}(u(\cdot, x)) = \sum_{i=0}^{N-1} |u(t_{i+1}, x) - u(t_i, x)|^{\frac{2\alpha}{\alpha - 1}}$$

$$\to_{N \to \infty} c_{2,\alpha}^{\frac{2\alpha}{\alpha - 1}} C_{\frac{1}{2}, 1 - \frac{1}{\alpha}}(A_2 - A_1)$$

in probability.

Proof. This is an immediate consequence of Proposition 5.16 and of Proposition 8.1. Proposition 5.16 states that

$$(u(t, x), t \geq 0) \equiv^{(d)} \left(c_{2,\alpha} B_t^{\frac{1}{2}, 1 - \frac{1}{\alpha}}, t \geq 0 \right)$$

with $c_{2,\alpha}$ defined in (5.49) and and $B^{\frac{1}{2}, 1 - \frac{1}{\alpha}}$ a bi-fBm with $H = \frac{1}{2}$ and $K = 1 - \frac{1}{\alpha}$ while Proposition 8.1 which gives the behavior of the q-variations of the bifractional Brownian motion $B^{\frac{1}{2}, 1 - \frac{1}{\alpha}}$. □

Remark 8.2. For $\alpha = 2$, we deduce that the solution to the standard heat equation with space-time white noise has non-trivial quartic variation, i.e.

$$S^{N,4}(u(\cdot, x)) = \sum_{i=0}^{N-1} |u(t_{i+1}, x) - u(t_i, x)|^4 \to_{N \to \infty} c_{2,2}^4 C_{\frac{1}{2}, \frac{1}{2}} (A_2 - A_1).$$

This was first noticed in [Swanson (2007)].

We defined the concept of renormalized q-variation for a perturbed fractional Brownian motion, see formula (8.7). By Proposition 5.16, the mild solution (5.46) is, modulo a constant, a perturbed fBm with Hurst index $H = \frac{1}{2}\left(1 - \frac{1}{\alpha}\right) = \frac{\alpha-1}{2\alpha} \in \left(0, \frac{1}{2}\right)$. We have the following.

Proposition 8.5. *Fix* $0 \leq A_1 < A_2$, *and* $x \in \mathbb{R}$. *Let* $t_j, j = 0, 1, \ldots, n$ *be the partition (2.14) of the interval* $[A_1, A_2]$. *Then for every* $x \in \mathbb{R}$ *and* $q \geq 2$ *integer,*

$$\frac{1}{\sqrt{n}} V_{q,n,[A_1,A_2]} \left(c_{2,\alpha}^{-1} 2^{-\frac{1}{2\alpha}} u(\cdot, x) \right)$$

$$= \frac{1}{\sqrt{n}} \sum_{i=0}^{n-1} \left[\left(\frac{n^{\frac{\alpha-1}{2\alpha}}}{c_{2,\alpha} 2^{\frac{1}{2\alpha}} (A_2 - A_1)^{\frac{\alpha-1}{2\alpha}}} \right)^q (u(t_{i+1}, x) - u(t_i, x))^q - \mu_q \right]$$

$$\to_{N \to \infty} N(0, \sigma_{\frac{1}{2}(1-\frac{1}{\alpha}),q}^2)$$

with $\sigma_{\frac{1}{2}(1-\frac{1}{\alpha}),q}^2$ *from (8.9).*

Proof. From Proposition 5.16, we know that

$$(u(t, x) + X_t, t \geq 0) =^{(d)} \left(c_{2,\alpha} 2^{\frac{1}{2\alpha}} B_t^{\frac{1}{2}(1-\frac{1}{\alpha})}, t \geq 0 \right)$$

where $B^{\frac{1}{2}(1-\frac{1}{\alpha})}$ is a fBm with Hurst parameter $\frac{1}{2}(1 - \frac{1}{\alpha})$ and $(X_t, t \geq 0)$ is a Gaussian process which has absolutely continuous paths. Therefore, $(c_{2,\alpha}^{-1} 2^{-\frac{1}{2\alpha}} u(t, x), t \geq 0)$ is a perturbed fBm in the sense of Lemma 8.1. Also note that its Hurst parameter is strictly less than $\frac{1}{2}$. We can then apply Lemma 8.2 to obtain the conclusion. □

8.3.2 Spatial variations

We let $t \geq 0$ be fixed and we look to the asymptotic behavior of the sequence (8.17). We suppose again $\alpha \in (1, 2]$ in order to include the case of the standard stochastic heat equation.

To get the behavior of the q-variations in space of the mild solution (5.46), we will use the decomposition obtained in Theorem 5.4 and the regularity of the process S_α defined by (5.55).

Proposition 8.6. *Fix $A_1 < A_2$ and $t > 0$. Let $n \geq 1$ and let $(x_j, j = 0, \ldots, n)$ be given by (8.18). Then the process $(u(t,x), x \in \mathbb{R})$ given by (5.46) has $\frac{2}{\alpha-1}$-variation, i.e. we have the following limit in probability, for every $t > 0$ fixed*

$$\lim_{n \to \infty} S_{[A_1, A_2]}^{n, \frac{2}{\alpha-1}}(u(t, \cdot)) = \lim_{n \to \infty} \sum_{j=0}^{n-1} |u(t, x_{j+1}) - u(t, x_j)|^{\frac{2}{\alpha-1}}$$

$$= \left(\sqrt{C_{0,\alpha}}\right)^{\frac{2}{\alpha-1}} \mathbf{E}|Z|^{\frac{2}{\alpha-1}}(A_2 - A_1)$$

with $C_{0,\alpha}$ from Theorem 5.4.

Proof. We recall the decomposition of u in Theorem 5.4. In Proposition 5.19 we showed that the process $(S_\alpha(t,x), x \in \mathbb{R})$ satisfies condition (8.3) (see the inequality (5.56)). We can then apply Lemma 8.1, point 1. to get the conclusion. □

With respect to the space variable, the process $(u(t,x), t \geq 0)$ is equal in distribution with $\sqrt{C_{0,\alpha}} B^{\frac{\alpha-1}{2}}$ plus a regular process, where $B^{\frac{\alpha-1}{2}}$ is a fBm with Hurst index $H = \frac{\alpha-1}{2} \in (0, \frac{1}{2}]$ and the constant $C_{0,\alpha}$ is that appearing in Theorem 5.4.

From Lemma 8.3 we have the following result.

Proposition 8.7. *Fix $A_1 < A_2$ and $t > 0$. Let $(x_j, j = 0, \ldots, n)$ be as in (8.18) with $n \geq 1$. Then if $\alpha \in (1, 2)$*

$$\frac{1}{\sqrt{n}} \sum_{i=0}^{n-1} \left[\left(\frac{n^{\frac{\alpha-1}{2}}}{\sqrt{C_{0,\alpha}}} \right)^q (u(t, x_{i+1}) - u(t, x_i))^q - \mu_q \right] \to N(0, \sigma_{\frac{\alpha-1}{2}, q}^2).$$

$$(8.20)$$

If $\alpha = 2$ (i.e. $\frac{\alpha-1}{2} = \frac{1}{2}$), then

$$\frac{1}{\sqrt{n}} \sum_{i=0}^{n-1} \left[\frac{n}{C_0} (u(t, x_{i+1}) - u(t, x_i))^2 - 1 \right] \to N(0, \sigma_{\frac{1}{2}, 2}^2)$$

with the constant $\sigma_{H,q}$ from (8.9) and C_0 from (5.35).

Proof. It suffices to note that, $\mathbf{E}|S_\alpha(x) - S_\alpha(y)|^2 \leq C|t - s|^2$, see relation (5.56), and that the Hurst index $\frac{\alpha-1}{2}$ is less than $\frac{1}{2}$. For $\alpha = 2$ the result is known from [Pospisil and and Tribe (2007)]. □

Let us make some comments:

Remark 8.3.

(1) When $\alpha = 2$, we have (exact) quadratic variation in space for the solution.
(2) We notice that the order q of a non-trivial q variation in time is α times the order of the spatial variation (i.e. $\frac{\alpha-1}{2\alpha}$-exact variation in time and $\frac{\alpha-1}{2}$-exact variation in space).

8.4 Variations of the solution to the heat equation with white-colored noise

Consider the mild solution $(u(t,x), t \geq 0, x \in \mathbb{R}^d)$ given by (6.20) and suppose that condition (6.21) holds true. Assume again $\alpha \in (1,2]$, the case $\alpha = 2$ corresponding to the mild solution to the standard heat equation with space-time white noise. We state the result concerning the limit of its q-temporal variation sequence by assuming that the space variable is fixed. The proofs are similar to those in the previous paragraph, the difference coming from the fact that we work on a higher spatial dimension and this leads to appearance of the isotropic fractional Brownian sheet.

Proposition 8.8. *Fix $A_1 < A_2$ and $x \in \mathbb{R}$. Let $t_j = A_1 + \frac{j}{n}(A_2 - A_1), n \geq 1, j = 0, 1, \ldots, n$ be a partition of the interval $[A_1, A_2]$. Then the process $(u(t,x), t \geq 0)$ given by (6.20) admits a variation of order $\frac{2\alpha}{\alpha+\gamma-d}$ which is equal to*

$$c_{2,\alpha,\gamma}^{\frac{2\alpha}{\alpha+\gamma-d}} C_{\frac{1}{2}, 1-\frac{d-\gamma}{\alpha}}(A_2 - A_1)$$

with $C_{\frac{1}{2}, 1-\frac{d-\gamma}{\alpha}}$ from (8.2) and $c_{2,\alpha,\gamma}$ from (6.26).

Proof. We have showed in Proposition 6.7 that $(u(t,x), t \geq 0)$ has the same finite dimensional distributions as $\left(c_{2,\alpha,\gamma} B_t^{\frac{1}{2}, 1-\frac{d-\gamma}{\alpha}}, t \geq 0\right)$ where $\left(B^{\frac{1}{2}, 1-\frac{d-\gamma}{\alpha}}, t \geq 0\right)$ is a bi-fBm with $H = \frac{1}{2}$ and $K = 1 - \frac{d-\gamma}{\alpha}$. The conclusion is a consequence of Corollary 8.1. $\square$

And from the proof of Proposition 8.5 and Lemma 8.2 we have

Proposition 8.9. *Fix $A_1 < A_2$ and $x \in \mathbb{R}$. Let $t_j = A_1 + \frac{j}{n}(A_2 - A_1), n \geq$*

$1, j = 0, 1, \ldots, n$ *be a partition of the interval* $[A_1, A_2]$. *Then*

$$\frac{1}{\sqrt{n}} \sum_{i=0}^{n-1} \left[\left(\frac{n^{\frac{\alpha+\gamma-d}{2\alpha}}}{c_{2,\alpha,\gamma} 2^{\frac{d-\gamma}{2\alpha}} (A_2 - A_1)^{\frac{\alpha+\gamma-d}{2\alpha}}} \right)^q (u(t_{i+1}, x) - u(t_i, x))^q - \mu_q \right]$$
$$\to^{(d)} N(0, \sigma^2_{\frac{1}{2}(1-\frac{d-\gamma}{\alpha}), q})$$

with $\sigma^2_{\frac{1}{2}(1-\frac{1}{\alpha}), q}$ *from (8.9) and* $c_{2,\alpha,\gamma}$ *from (6.26).*

When $\gamma = 0$ and $d = 1$, we retrieve (modulo a constant) the result in the case of the white noise in space.

For the spatial variation, we need to recall the decomposition of the solution from Theorem 6.2. This results says that the process $(u(t, x), x \in \mathbb{R}^d)$ with $t > 0$ fixed, is a perturbed isotropic fractional Brownian motion. We also recall Lemma 8.6, where we stated the limit behavior of the (exact and renormalized) perturbed isotropic fractional Brownian motion. From Lemma 8.6 and Theorem 6.2, we have

Proposition 8.10. *The process* $x \to u(t, x)$ *has* $\frac{2}{\alpha+\gamma-d}$-*variation given by*

$$\sqrt{C_{0,\alpha,\gamma}}^{-\frac{2}{\alpha+\gamma-d}} (A_2 - A_1) d\mathbf{E}|Z|^{\frac{2}{\alpha+\gamma-d}}$$

with $C_{0,\alpha,\gamma}$ *from (6.31).*

and

Proposition 8.11. *Fix* $A_1 < A_2$ *and* $t > 0$. *Let* $x_j^k = A_1 + \frac{j}{n}(A_2 - A_1)$ *for* $j = 0, \ldots, n$, $n \geq 1$ *and for every* $k = 1, \ldots, d$. *Also let* $\mathbf{x}_j = (x_j^1, \ldots, x_j^d)$. *Then*

$$\frac{1}{\sqrt{n}} \sum_{i=0}^{n-1} \left[\left(\frac{n^{\frac{\alpha+\gamma-d}{2}} d^{-H}}{\sqrt{C_{0,\alpha,\gamma}}} \right)^q (u(t, \mathbf{x}_{i+1}) - u(t, \mathbf{x}_i))^q - \mu_q \right]$$
$$\to_{n\to\infty} N(0, \sigma^2_{\frac{\alpha+\gamma-d}{2}, q})$$

with $C_{0,\alpha,\gamma}$ *from (6.31) and* $\sigma_{\frac{\alpha+\gamma-d}{2}, q}$ *from (8.20).*

Notice that we used the fact that $0 < \frac{1}{2}\left(1 - \frac{d-\gamma}{\alpha}\right) < \frac{1}{2}$ which means that the parameter of the isotropic fBm associated to the solution is strictly less than one-half and we can apply Lemma 8.6.

Chapter 9

Parameter estimation for the stochastic heat equation via power variations

The p-variations are very useful in order to identify the various parameters of stochastic processes in general and of solutions to stochastic equations in particular. Some examples are presented in e.g. the monograph [Tudor (2013)]. More recently, this method started to be intensively used to do statistical inference for stochastic partial differential equations. We will illustrate this method for the case of the stochastic heat equation (with white or colored noise in space and with white noise in time) and for the stochastic wave equation driven by the space-time white noise. An obvious observation is that, based on p-variations, we can immediately construct good estimators for the diffusion (or volatility) parameter of these equations. To exemplify, consider the usual stochastic heat equation

$$\frac{\partial u}{\partial t}(t, x) = \Delta u(t, x) + \sigma \dot{W}(t, x), \quad t \geq 0, x \in \mathbb{R} \tag{9.1}$$

with W the space-time white noise and with vanishing initial value $u(0, x) = 0$ for every $x \in \mathbb{R}$. The purpose is to estimate the parameter $\sigma > 0$ in (9.1) based on the discrete observations of the solution u in time or in space. That is, we suppose that we dispose on the data $(u(t_i, x), i = 0, \ldots, N)$ with $t_i, i = 0, 1, \ldots, N$ given by (2.14) and for some $x \in \mathbb{R}$ or we have at our disposal the observations $(u(t, x_i), i = 0, \ldots, N)$ with $t > 0$ fixed and $x_i, i = 0, 1, \ldots, N$ given by (8.18). It is obvious that if we consider the temporal quartic sequence given by (8.16)

$$S_{[A_1, A_2]}^{N,4}(u(\cdot, x)) = \sum_{i=0}^{N-1} |u(t_{i+1}, x) - u(t_i, x)|^4$$

then by Proposition 8.4, in probability,

$$S_{[A_1, A_2]}^{N,4}(u(\cdot, x)) \to_{N \to \infty} K_2 \sigma^4$$

with $K_2 = c_{2,2}^4 C_{\frac{1}{2},\frac{1}{2}}(A_2 - A_1)$ with $C_{1/2,\frac{1}{2}}$ from (8.2) and $c_{2,2}$ from (5.49) with $\alpha = 2$. Therefore the quartic variation

$$K_2^{-1} S_{[A_1,A_2]}^{N,4}(u(\cdot,x))$$

constitutes a consistent estimator the the parameter σ^4, i.e. this sequence converges to σ^4 in probability. Moreover, Proposition 8.5 indicates that the sequence

$$K_2^{-1} S_{[A_1,A_2]}^{N,4}(u(\cdot,x)) - \sigma^4$$

satisfies a CLT, after a proper renormalization. Similarly, one can use the spatial data $(u(t,x_i), i = 0, 1, \ldots, N)$ with $x_i, i = 0, 1, \ldots, N$ in (8.18) and the spatial variation sequence (8.17) in order to get a consistent and asymptotically normal estimator for σ^2, based on the limit theorems proven in Section 8.3.2. Moreover, this approach can be applied to the fractional stochastic heat equation with space-time white noise and to the (fractional) stochastic heat equation with white-colored Gaussian noise.

Below we will illustrate how the p-variations method can be used for statistical inference for the drift parameter of the heat equation.

9.1 The parametrized heat equation with space-time white noise

We consider the parametrized stochastic heat equation

$$\frac{\partial u_\theta}{\partial t}(t,x) = -\theta(-\Delta)^{\frac{\alpha}{2}} u_\theta(t,x) + \dot{W}(t,x), \quad t \geq 0, x \in \mathbb{R} \qquad (9.2)$$

with vanishing initial conditions, where $(-\Delta)^{\frac{\alpha}{2}}$ denotes the fractional Laplacian of order $\alpha \in (1,2]$, $\theta > 0$ and W is a Gaussian noise which is white and in space, i.e. its covariance function is given by (5.4). In the particular case $\alpha = 2$ (which will be included in our analysis below), (9.2) reads

$$\frac{\partial u_\theta}{\partial t}(t,x) = \theta \Delta u(t,x) + \dot{W}(t,x), \quad t \geq 0, x \in \mathbb{R}.$$

The purpose is to estimate the drift parameter $\theta > 0$ based on the observation of the solution to (9.2) at discrete times or at discrete points in space.

9.1.1   *Properties of the solution*

Let us first discuss the properties of the solution to the above SPDE. Let $G_\alpha(t, x)$ be the Green kernel associated to the operator $-(-\Delta)^{\frac{\alpha}{2}}$, see Section 5.2. Then the Green kernel associated to the operator operator $-\theta(-\Delta)^{\frac{\alpha}{2}}$ is

$$G_\alpha(\theta t, x).$$

This means the the the kernel $H_\alpha(t, x) = G_\alpha(\theta t, x)$ solves $\frac{\partial H_\alpha}{\partial t}(t, x) = -\theta(-\Delta)^{\frac{\alpha}{2}} H_\alpha(t, x)$. Therefore, the mild solution to (9.2) can be expressed as

$$u_\theta(t, x) = \int_0^t \int_{\mathbb{R}} G_\alpha(\theta(t - s), x - y) W(ds, dy) \tag{9.3}$$

where the stochastic integral $W(ds, dy)$ is an usual Wiener integral with respect to the space-time white noise, see Section 4.4. For $\theta = 1$, the solution to the heat equation (9.2) coincides with the random field (5.46).

We start by consider a certain transform of the solution (9.3).

Lemma 9.1. *Suppose that the process $(u_\theta(t, x), t \geq 0, x \in \mathbb{R})$ satisfies (9.2) in the mild sense. Define*

$$v_\theta(t, x) = u_\theta\left(\frac{t}{\theta}, x\right), \quad t \geq 0, x \in \mathbb{R}. \tag{9.4}$$

Then the process $(v_\theta(t, x), t \geq 0, x \in \mathbb{R})$ satisfies

$$\frac{\partial v_\theta}{\partial t}(t, x) = -(-\Delta)^{\frac{\alpha}{2}} v_\theta(t, x) + (\theta)^{-\frac{1}{2}} \dot{\tilde{W}}(t, x), \quad t \geq 0, x \in \mathbb{R} \tag{9.5}$$

with $v_\theta(0, x) = 0$ for every $x \in \mathbb{R}$, where $\tilde{W}$ is a space-time white noise, i.e a centered Gaussian random field with covariance (3.5). More exactly, we have for every $t \geq 0, x \in \mathbb{R}$,

$$v_\theta(t, x) = \theta^{-\frac{1}{2}} \int_0^t \int_{\mathbb{R}} G_\alpha(t - s, x - y) \tilde{W}(ds, dy). \tag{9.6}$$

Proof. From (9.3), we have for every $t \geq 0, x \in \mathbb{R}$,

$$v_\theta(t, x) = u_\theta\left(\frac{t}{\theta}, x\right) = \int_0^{\frac{t}{\theta}} \int_{\mathbb{R}} G_\alpha(t - \theta s, x - y) W(ds, dy)$$

$$= \int_0^t \int_{\mathbb{R}} G_\alpha(t - s, x - y) W(d\frac{s}{\theta}, dy)$$

$$= \theta^{-\frac{1}{2}} \int_0^t \int_{\mathbb{R}} G_\alpha(t - s, x - y) \tilde{W}(ds, dy)$$

where, for $t \geq 0, A \in \mathcal{B}(\mathbb{R})$, we denoted $\tilde{W}(t, A) = \theta^{\frac{1}{2}} W\left(\frac{t}{\theta}, A\right)$. Notice that $\tilde{W}$ has the same finite dimensional distributions as W, due to the scaling property of the white noise. $\square$

We can deduce the law of the process u_θ in time and space.

Proposition 9.1. *Let $(u_\theta(t, x), t \geq 0, x \in \mathbb{R})$ be given by (9.3). For every $x \in \mathbb{R}$ and $\theta > 0$, we have*

$$(u_\theta(t, x), t \geq 0) \equiv^{(d)} \left(\theta^{-\frac{1}{2\alpha}} c_{2,\alpha} B_t^{\frac{1}{2}, 1-\frac{1}{\alpha}}, t \geq 0 \right)$$

where

$$\left(B^{\frac{1}{2}, 1-\frac{1}{\alpha}}, t \geq 0 \right)$$

is a bifractional Brownian motion with parameters $H = \frac{1}{2}$ and $K = 1 - \frac{1}{\alpha}$ and $c_{2,\alpha}$ is given by (5.49).

Proof. Fix $x \in \mathbb{R}$ and $\theta > 0$. Then for every $s, t \geq 0$, we have by (9.6) and Proposition 5.16,

$$\begin{aligned}
&\mathbf{E} u_\theta(t, x) u_\theta(s, x) \\
&= \mathbf{E} v_\theta(\theta t, x) v_\theta(\theta s, x) \\
&= \theta^{-1} \mathbf{E} u_1(\theta t, x) u_1(\theta s, x) = \theta^{-1} c_{1,\alpha} \left[(\theta t + \theta s)^{1-\frac{1}{\alpha}} - |\theta t - \theta s|^{1-\frac{1}{\alpha}} \right] \\
&= \theta^{-\frac{1}{\alpha}} c_{2,\alpha}^2 \mathbf{E} B_t^{\frac{1}{2}, 1-\frac{1}{\alpha}} B_s^{\frac{1}{2}, 1-\frac{1}{\alpha}}
\end{aligned}$$

with $c_{1,\alpha}$ from Proposition 5.16 and with $B^{\frac{1}{2}, 1-\frac{1}{\alpha}}$ a bifractional Brownian motion with parameters $H = \frac{1}{2}$ and $K = 1 - \frac{1}{\alpha}$

     $\square$

Concerning the spatial distribution of the solution, we have the following result.

Proposition 9.2. *For every $t \geq 0, \theta > 0$, we have the following equality in distribution*

$$(u_\theta(t, x), x \in \mathbb{R}) \equiv^{(d)} \left(\theta^{-\frac{1}{2}} \sqrt{C_{0,\alpha}} B^{\frac{\alpha-1}{2}}(x) + S_{\theta t}(x), x \in \mathbb{R} \right)$$

where $B^{\frac{\alpha-1}{2}}$ is a fractional Brownian motion with Hurst parameter $\frac{\alpha-1}{2} \in (0, \frac{1}{2}]$, $(S_{\theta t}(x))_{x \in \mathbb{R}}$ is a centered Gaussian process with C^∞ sample paths and $C_{0,\alpha}$ from (5.58).

Proof. The result is immediate since for every $t > 0, \theta > 0$

$$(u_\theta(t,x), x \in \mathbb{R}) = (v_\theta(\theta t, x), x \in \mathbb{R}) \equiv^{(d)} \theta^{-\frac{1}{2}} (u_1(t,x), x \in \mathbb{R})$$

$$\equiv^{(d)} \left(\theta^{-\frac{1}{2}} \sqrt{C_{0,\alpha}} B^{\frac{\alpha-1}{2}}(x) + S_{\theta t}(x), x \in \mathbb{R}\right)$$

where we used Theorem 5.4.   □

Notice that the Hurst parameter $\frac{\alpha-1}{2}$ of the fBm in Proposition 9.2 belongs to the interval $(0, \frac{1}{2}]$; it may be $\frac{1}{2}$ if $\alpha = 2$.

Therefore, by using again the notation $t_j = A_1 + \frac{j}{n}(A_2 - A_1), n \geq 1, j = 0, \dots, n$ (see 2.14), we have via Proposition 8.4, the following limit theorem.

Lemma 9.2. *Let u_θ be the solution to (9.2). Then for every $x \in \mathbb{R}$*

$$S^{n, \frac{2\alpha}{\alpha-1}}_{[A_1, A_2]}(u_\theta(\cdot, x)) := \sum_{j=0}^{n-1} |u_\theta(t_{j+1}, x) - u_\theta(t_j, x)|^{\frac{2\alpha}{\alpha-1}} \tag{9.7}$$

$$\to_{n \to \infty} c_{2,\alpha}^{\frac{2\alpha}{\alpha-1}} C_{\frac{1}{2}, 1-\frac{1}{\alpha}}(A_2 - A_1)|\theta|^{\frac{-1}{\alpha-1}}$$

$$= c_{2,\alpha}^{\frac{2\alpha}{\alpha-1}} 2^{\frac{1}{\alpha-1}} \mu_{\frac{2\alpha}{\alpha-1}}(A_2 - A_1)|\theta|^{\frac{-1}{\alpha-1}}$$

in probability.

Proof. We have by (9.4) and (9.6) (the limit below is in probability)

$$\lim_{n \to \infty} \sum_{j=0}^{n-1} |u_\theta(t_{j+1}, x) - u_\theta(t_j, x)|^{\frac{2\alpha}{\alpha-1}}$$

$$= \lim_{n \to \infty} \sum_{j=0}^{n-1} |v_\theta(\theta t_{j+1}, x) - u_\theta(\theta t_j, x)|^{\frac{2\alpha}{\alpha-1}}$$

$$= \theta^{-\frac{\alpha}{\alpha-1}} \sum_{j=0}^{n-1} |u_1(\theta t_{j+1}, x) - u_1(\theta t_j, x)|^{\frac{2\alpha}{\alpha-1}}$$

$$= \theta^{-\frac{\alpha}{\alpha-1}} c_{2,\alpha}^{\frac{2\alpha}{\alpha-1}} C_{\frac{1}{2}, 1-\frac{1}{\alpha}}(\theta A_2 - \theta A_1) = c_{2,\alpha}^{\frac{2\alpha}{\alpha-1}} C_{\frac{1}{2}, 1-\frac{1}{\alpha}}(A_2 - A_1)|\theta|^{\frac{-1}{\alpha-1}},$$

where we used the fact that $(\theta t_i, i = 0, \dots, n)$ constitutes a partition of ther interval $[\theta A_1, \theta A_2]$.   □

The result in Proposition 9.2 says that the process $\theta^{\frac{1}{2}}(\sqrt{C_{0,\alpha}})^{-1}(u_\theta(t,x), x \in \mathbb{R})$ is a perturbed fBm, so we know its exact variation in space. Below $x_j = A_1 + \frac{j}{n}(A_2 - A_1), j = 0, \dots, n$ will denote the partition (8.18) of the interval $[A_1, A_2]$.

Proposition 9.3. *Let u_θ be given by (9.2). Then, with $C_{0,\alpha}$ from (5.58),*

$$\sum_{i=0}^{n-1} |u_\theta(t, x_{j+1}) - u_\theta(t, x_j)|^{\frac{2}{\alpha-1}} \to_{n\to\infty} \sqrt{C_{0,\alpha}}^{\frac{-2}{\alpha-1}} \mu_{\frac{2}{\alpha-1}} (A_2 - A_1) |\theta|^{\frac{-1}{\alpha-1}}$$

and if $q := \frac{2}{\alpha-1}$ is an integer

$$\frac{1}{\sqrt{n}} \sum_{i=0}^{n-1} \left[\left(\frac{n}{\sqrt{C_{0,\alpha}}^{\frac{2}{\alpha-1}} (A_2 - A_1)} \right)^q \theta^{\frac{1}{\alpha-1}} (u_\theta(t, x_{i+1}) - u_\theta(t, x_i))^{\frac{2}{\alpha-1}} - \mu_{\frac{2}{\alpha-1}} \right]$$

$$\to^{(d)} N(0, \sigma^2_{\frac{\alpha-1}{2}, \frac{2}{\alpha-1}})$$

9.1.2 *Estimators of the drift parameter*

Relation (9.7) motivates the definition of the following estimator for the parameter $\theta > 0$ of the model (9.2)

$$\widehat{\theta}_{n,1} = \left(\left(c_{2,\alpha}^{\frac{2\alpha}{\alpha-1}} 2^{\frac{1}{\alpha-1}} \mu_{\frac{2\alpha}{\alpha-1}} (A_2 - A_1) \right)^{-1} \right.$$

$$\left. \times \sum_{i=0}^{n-1} |u_\theta(t_{j+1}, x) - u_\theta(t_j, x)|^{\frac{2\alpha}{\alpha-1}} \right)^{1-\alpha} \tag{9.8}$$

$$= \left(c_{2,\alpha}^{\frac{2\alpha}{\alpha-1}} 2^{\frac{1}{\alpha-1}} \mu_{\frac{2\alpha}{\alpha-1}} (A_2 - A_1) \right)^{\alpha-1} \left(S^{n, \frac{2\alpha}{\alpha-1}} (u_\theta(\cdot, x)) \right)^{1-\alpha}$$

and so

$$\widehat{\theta}_{n,1}^{\frac{1}{1-\alpha}} = \frac{1}{c_{2,\alpha}^{\frac{2\alpha}{\alpha-1}} 2^{\frac{1}{\alpha-1}} \mu_{\frac{2\alpha}{\alpha-1}} (A_2 - A_1)} \sum_{i=0}^{n-1} |u_\theta(t_{j+1}, x) - u_\theta(t_j, x)|^{\frac{2\alpha}{\alpha-1}}. \tag{9.9}$$

In order to have the above quantity well-defined we need to assume that $\frac{2\alpha}{\alpha-1}$ is an even integer. From the results in Chapter 8, we obtain the strong consistency and the asymptotic normality of the above estimator.

Proposition 9.4. *Assume $\frac{2\alpha}{\alpha-1} := q$ is an even integer and consider the estimator $\widehat{\theta}_{n,1}$ defined by (9.8). Then $\widehat{\theta}_{n,1} \to_{n\to\infty} \theta$ in probability and*

$$\sqrt{n} \left[\widehat{\theta}_{n,1}^{\frac{1}{1-\alpha}} - \theta^{\frac{1}{1-\alpha}} \right] \to^{(d)} N(0, s^2_{1,\theta,\alpha}) \text{ with } s^2_{1,\theta,\alpha} = \sigma^2_{\frac{1}{q}, q} \theta^{\frac{2}{1-\alpha}} \mu^{-2}_{\frac{2\alpha}{\alpha-1}}. \tag{9.10}$$

Proof. The consistency of a direct consequence of Lemma 9.2. From Proposition 9.1 and relation between the fBm and the bi-fBm (Theorem 2.2), we obtain that

$$\left(u_\theta(t,x) + c_{2,\alpha}\theta^{-\frac{1}{2\alpha}} X_t \right) \equiv^{(d)} c_{2,\alpha}\theta^{-\frac{1}{2\alpha}} 2^{\frac{1}{2\alpha}} B^{\frac{\alpha-1}{2\alpha}}$$

where $B^{\frac{\alpha-1}{2\alpha}}$ is a fBm with Hurst parameter $\frac{\alpha-1}{2\alpha} \in (0, \frac{1}{2})$. Therefore, $\left(c_{2,\alpha}\theta^{-\frac{1}{2\alpha}} 2^{\frac{1}{2\alpha}} \right)^{-1} u_\theta$ is a perturbed fBm and we obtain, by taking $H = \frac{\alpha-1}{2\alpha}$ and $q = \frac{1}{H} = \frac{2\alpha}{\alpha-1}$ in Lemma 8.5

$$\frac{1}{\sqrt{n}} \sum_{i=0}^{n-1} \left[\frac{n\theta^{\frac{1}{\alpha-1}}}{c_{2,\alpha}^{\frac{2\alpha}{\alpha-1}} 2^{\frac{1}{\alpha-1}} (A_2 - A_1)} \left(u_\theta(t_{j+1},x) - u_\theta(t_j,x) \right)^{\frac{2\alpha}{\alpha-1}} - \mu_{\frac{2\alpha}{\alpha-1}} \right]$$
$$\to^{(d)} N(0, \sigma^2_{\frac{1}{q},q}).$$

Or, this means, due to the expression (10.23) of the estimator $\widehat{\theta}_{n,1}$,

$$\sqrt{n}\mu_{\frac{2\alpha}{\alpha-1}} \theta^{\frac{1}{\alpha-1}} \left[\widehat{\theta}_{n,1}^{\frac{1}{1-\alpha}} - \theta^{\frac{1}{1-\alpha}} \right] \to^{(d)} N(0, \sigma^2_{\frac{1}{q},q})$$

which is equivalent to (9.10). $\square$

Using the so-called delta-method (see e.g. [Asmussen (2005)]), we can get the asymptotic behavior of the estimator $\widehat{\theta}_n$. Recall that if $(X_n)_{n\geq 1}$ is a sequence of random variables such that

$$\sqrt{n}(X_n - \gamma_0) \to^{(d)} N(0, \sigma^2)$$

and g is a function such that $g'(\gamma_0)$ exists and does not vanish, then

$$\sqrt{n}(g(X_n) - g(\gamma_0)) \to^{(d)} N(0, \sigma^2 g'(\gamma_0)^2). \tag{9.11}$$

Proposition 9.5. *Consider the estimator (9.8) and let $s_{1,\theta,\alpha}$ be given by (9.10). Then as $n \to \infty$*

$$\sqrt{n}(\widehat{\theta}_{n,1} - \theta) \to N(0, s^2_{1,\theta,\alpha}(1-\alpha)^2 \theta^{\frac{2\alpha}{\alpha-1}}). \tag{9.12}$$

Proof. By applying the delta-method for the function $g(x) = x^{1-\alpha}$, $X_n = \widehat{\theta}_{n,1}^{\frac{1}{1-\alpha}}$ and $\gamma_0 = \theta^{\frac{1}{1-\alpha}}$, we immediately obtain the convergence (9.12).

$\square$

It is possible to define an estimator for the parameter θ based on the spatial variations of the solution (9.3).

The above Proposition 9.3 leads to the definition of the estimator

$$\widehat{\theta}_{n,2} = \left[(C_{0,\alpha}^{\frac{2}{\alpha-1}} \mu_{\frac{2}{\alpha-1}} (A_2 - A_1))^{-1} \sum_{i=0}^{n-1} |u_\theta(t,x_{j+1}) - u_\theta(t,x_j)|^{\frac{2}{\alpha-1}} \right]^{1-\alpha} \tag{9.13}$$

and we can immediately deduce from Proposition 9.2 its asymptotic properties.

Proposition 9.6. *The estimator (9.13) converges in probability as $n \to \infty$ to the parameter θ. Moreover, if $q := \frac{2}{\alpha-1}$ is an even integer*

$$\sqrt{n}\left[\widehat{\theta}_{n,2}^{\frac{1}{1-\alpha}} - \theta^{\frac{1}{1-\alpha}}\right] \to^{(d)} N(0, s_{2,\theta,\alpha}^2) \qquad (9.14)$$

with

$$s_{2,\theta,\alpha}^2 = \sigma_{\frac{\alpha-1}{2},\frac{2}{\alpha-1}}^2 \mu_{\frac{2}{\alpha-1}}^{-2} \theta^{\frac{2}{1-\alpha}}.$$

Proof. Using the law of the process $(u_\theta(t,x), x \in \mathbb{R})$ obtained in Proposition 9.2, we deduce that the Gaussian process $\left(\theta^{\frac{1}{2}}\sqrt{C_{0,\alpha}}^{-1} u_\theta(t,x), x \in \mathbb{R}\right)$ is a perturbed fractional Brownian motion. Therefore, by relation (8.9) in Lemma 8.3,

$$\frac{1}{\sqrt{n}} \sum_{i=0}^{n-1} \left(\frac{n\theta^{\frac{1}{\alpha-1}}}{(A_2 - A_1)\sqrt{C_{0,\alpha}}^{\frac{2}{\alpha-1}}} (u_\theta(t,x_{j+1}) - u_\theta(t,x_j))^{\frac{2}{\alpha-1}} - \mu_{\frac{2}{\alpha-1}} \right)$$

$$= \sqrt{n}\mu_{\frac{2}{\alpha-1}}\theta^{\frac{1}{\alpha-1}}\left[\widehat{\theta}_{n,2}^{\frac{1}{1-\alpha}} - \theta^{\frac{1}{1-\alpha}}\right] \to_{n\to\infty}^{(d)} N\left(0, \sigma_{\frac{\alpha-1}{2},\frac{2}{\alpha-1}}^2\right).$$

$\square$

By using the delta-method, we can obtain the asymptotic distribution of $\widehat{\theta}_{n,2}$.

Proposition 9.7. *Let $\widehat{\theta}_{n,2}$ be given by (9.13). Then, with $s_{2,\theta,\alpha}$ from (9.14), as $n \to \infty$*

$$\sqrt{n}(\widehat{\theta}_{n,2} - \theta) \to^{(d)} N\left(0, s_{2,\theta,\alpha}^2(1-\alpha)^2\theta^{\frac{2\alpha}{\alpha-1}}\right).$$

Proof. It suffices to apply (9.11) to the function $g(x) = x^{1-\alpha}$ and $\gamma_0 = \theta^{\frac{1}{1-\alpha}}$ and to follow the proof of Proposition 9.5. $\square$

9.2 The parametrized heat equation with correlated spatial noise

Next we will consider the stochastic heat equation with an additive Gaussian noise which behaves as a Wiener process in time and as a fractional Brownian motion in space, i.e. its spatial covariance is given by the so-called Riesz kernel. We will again study the distribution of the solution, its connection with fractional and bifractional Brownian motion and we apply the q-variation method to obtain an asymptotically normal estimator for the drift parameter.

We will consider the stochastic heat equation

$$\frac{\partial}{\partial t} u_\theta(t, x) = -\theta(-\Delta)^{\frac{\alpha}{2}} u_\theta(t, x) + \dot{W}^\gamma(t, x), \qquad t \geq 0, x \in \mathbb{R}^d. \qquad (9.15)$$

with $u_\theta(0, x) = 0$ for every $x \in \mathbb{R}^d$. In (9.15), $-(-\Delta)^{\frac{\alpha}{2}}$ denotes the fractional Laplacian with exponent $\frac{\alpha}{2}$, $\alpha \in (1, 2]$ and W^γ is the so-called white-colored noise, i.e. $\left(W^\gamma(t, A), t \geq 0, A \in \mathcal{B}(\mathbb{R}^d)\right)$ is a centered Gaussian field with covariance

$$\mathbf{E} W^\gamma(t, A) W^\gamma(s, B) = (t \wedge s) \int_A \int_B f(x - y) dx dy \qquad (9.16)$$

where f is the Riesz kernel of order γ given by (3.13). As usual, the mild solution to (9.15) is given by, for $t \geq 0, x \in \mathbb{R}^d$,

$$u_\theta(t, x) = \int_0^t \int_{\mathbb{R}^d} G_\alpha(\theta(t - s), x - z) W^\gamma(ds, dz) \qquad (9.17)$$

where the above integral $W^\gamma(ds, dz)$ is a Wiener integral with respect to the Gaussian noise W^γ.

We know that, when $\theta = 1$, the mild solution (9.17) is well-defined as a square integrable process if and only if (6.21) holds, i.e. $d < \gamma + \alpha$. Moreover, for every $T > 0$,

$$\sup_{t \in [0,T], x \in \mathbb{R}^d} \mathbf{E} |u_1(t, x)|^2 < \infty.$$

In particular, condition (6.21) shows that the solution exists in any spatial dimension d, via a suitable choice of the parameter γ.

9.2.1 *Properties of the solution*

Recall also the following facts:

- Assume (6.21) is satisfied. Then, by Proposition 6.8, for every $x \in \mathbb{R}^d$, the we have the following equivalence in distribution

$$\left(u_1(t, x), t \geq 0\right) \equiv^{(d)} \left(c_{2,\alpha,\gamma} B_t^{\frac{1}{2}, 1 - \frac{d-\gamma}{\alpha}}, t \geq 0\right) \qquad (9.18)$$

  where $B^{\frac{1}{2}, 1 - \frac{d-\gamma}{\alpha}}$ is a bifractional Brownian motion with Hurst parameters $H = \frac{1}{2}$ and $K = 1 - \frac{d-\gamma}{\alpha}$ and $c_{2,\alpha,\gamma}$ defined by (6.26)

- For every $t \geq 0$, we have, by Theorem 6.2,

$$\left(u(t,x), x \in \mathbb{R}^d \right) \equiv^{(d)} \left(\sqrt{C_{0,\alpha,\gamma}} B^{\frac{\alpha+\gamma-d}{2}}(x) + S_t(x), x \in \mathbb{R}^d \right) \quad (9.19)$$

where $B^{\frac{\alpha+\gamma-d}{2}}$ is an isotropic d-dimensional fractional Brownian motion (see the next section) with Hurst parameter $\frac{\alpha+\gamma-d}{2}$, $(S_t(x), x \in \mathbb{R}^d)$ is a centered Gaussian process with C^∞ sample paths and $C_{0,\alpha,\gamma}$ is an explicit numerical constant, see (6.31).

Let us start by analyzing the distribution of the solution to (9.15) and its link with the (bi)fractional Brownian motion.

Proposition 9.8. *For every $x \in \mathbb{R}^d$ and $\theta > 0$, we have*

$$(u_\theta(t,x), t \geq 0) \equiv^{(d)} \left(\theta^{-\frac{d-\gamma}{2\alpha}} c_{2,\alpha,\gamma} B_t^{\frac{1}{2}, 1 - \frac{d-\gamma}{\alpha}}, t \geq 0 \right)$$

where $B^{\frac{1}{2}, 1 - \frac{d-\gamma}{\alpha}}$ is a bifractional Brownian motion with parameters $H = \frac{1}{2}$ and $K = 1 - \frac{d-\gamma}{\alpha}$ and the constant $c_{2,\alpha,\gamma}$ is defined by (6.26).

Proof. Denote

$$v_\theta(t,x) = u_\theta\left(\frac{t}{\theta}, x\right) \quad \text{for every } t \geq 0, x \in \mathbb{R}^d.$$

Then, as in Lemma 9.1, v_θ solves the equation

$$\frac{\partial v_\theta}{\partial t}(t,x) = -(-\Delta)^{\frac{\alpha}{2}} v_\theta(t,x) + \theta^{-\frac{1}{2}} \dot{\widetilde{W}}^\gamma(t,x), \quad t \geq 0, x \in \mathbb{R} \quad (9.20)$$

with $v_\theta(0,x) = 0$ for every $x \in \mathbb{R}$, where $\dot{\widetilde{W}}^\gamma$ is a white colored Gaussian noise (i.e. a Gaussian process with zero mean and covariance (9.16)). That means that for every $t \geq 0, x \in \mathbb{R}^d$,

$$v_\theta(t,x) = \theta^{-\frac{1}{2}} \int_0^t \int_{\mathbb{R}^d} G_\alpha(t-s, x-y) \widetilde{W}^\gamma(ds, dy). \quad (9.21)$$

Fix $x \in \mathbb{R}^d$ and $\theta > 0$. We have, for every $s, t \geq 0$, we have by Proposition 6.2 and (9.21),

$$\begin{aligned}
\mathbf{E} u_\theta(t,x) u_\theta(s,x) &= \mathbf{E} v_\theta(\theta t, x) v_\theta(\theta s, x) \\
&= \theta^{-1} \mathbf{E} u_1(\theta t, x) u_1(\theta s, x) \\
&= \theta^{-1} c_{1,\alpha,\gamma} \left[(\theta t + \theta s)^{1 - \frac{d-\gamma}{\alpha}} - |\theta t - \theta s|^{1 - \frac{d-\gamma}{\alpha}} \right] \\
&= \theta^{-\frac{d-\gamma}{\alpha}} c_{2,\alpha,\gamma}^2 \mathbf{E} B_t^{\frac{1}{2}, 1 - \frac{d-\gamma}{\alpha}} B_s^{\frac{1}{2}, 1 - \frac{d-\gamma}{\alpha}}.
\end{aligned}$$

$\square$

For the behavior with respect to the space variable, we obtain the following result.

Proposition 9.9. *For every $t \geq 0, \theta > 0$, we have the following equality in distribution*

$$\left(u_\theta(t,x), x \in \mathbb{R}^d\right) \stackrel{(d)}{=} \left(\theta^{-\frac{1}{2}}\sqrt{C_{0,\alpha,\gamma}}B^{\frac{\alpha+\gamma-d}{2}}(x) + S_{\theta t}(x), x \in \mathbb{R}^d\right)$$

where $B^{\frac{\alpha-1}{2}}$ is an isotropic fractional Brownian sheet with Hurst parameter $\frac{\alpha-1}{2} \in (0, \frac{1}{2}]$, $(S_{\theta t}(x))_{x \in \mathbb{R}}$ is a centered Gaussian process with C^∞ sample paths and $C_{0,\alpha,\gamma}$ from (6.31).

Proof. The result is immediate since for fixed time $t > 0$

$$\left(u_\theta(t,x), x \in \mathbb{R}^d\right) = \left(v_\theta(\theta t, x), x \in \mathbb{R}^d\right) \stackrel{(d)}{=} \theta^{-\frac{1}{2}}\left(u_1(\theta t, x), x \in \mathbb{R}^d\right)$$

$$\stackrel{(d)}{=} \left(\theta^{-\frac{1}{2}}\sqrt{C_{0,\alpha,\gamma}}B^{\frac{\alpha+\gamma-d}{2}}(x) + S_{\theta t}(x), x \in \mathbb{R}^d\right).$$

$\square$

Again $t_j = A_1 + \frac{j}{n}(A_2 - A_1), j = 0, \ldots, n$ will denote a partition of the interval $[A_1, A_2]$.

Lemma 9.3. *Assume (3.8). Let u_θ be the solution to (9.15). Then for every $x \in \mathbb{R}^d$, the process $(u_\theta(t,x), t \geq 0)$ admits $\frac{2\alpha}{\alpha+\gamma-d}$-variation over the interval $[A_1, A_2]$, i.e.*

$$S_{[A_1,A_2]}^{n,\frac{2\alpha}{\alpha+\gamma-d}}(u(\cdot,x)) := \sum_{j=0}^{n-1}|u_\theta(t_{j+1},x) - u_\theta(t_j,x)|^{\frac{2\alpha}{\alpha+\gamma-d}} \quad (9.22)$$

$$\to_{n\to\infty} c_{2,\alpha,\gamma}^{\frac{2\alpha}{\alpha+\gamma-d}}2^{\frac{d-\gamma}{\alpha+\gamma-d}}\mu_{\frac{2\alpha}{\alpha+\gamma-d}}(A_2 - A_1)|\theta|^{\frac{\gamma-d}{\alpha+\gamma-d}}$$

in probability.

Proof. Clearly, for fixed $x \in \mathbb{R}^d$,

$$\sum_{i=0}^{n-1}|u_\theta(t_{j+1},x) - u_\theta(t_j,x)|^{\frac{2\alpha}{\alpha+\gamma-d}} = \sum_{i=0}^{n-1}|v(\theta t_{j+1},x) - v(\theta t_j,x)|^{\frac{2\alpha}{\alpha+\gamma-d}}$$

where $(v_\theta(t,x), t \geq 0) \stackrel{(d)}{=} (\theta^{-\frac{1}{2}}u_1(t,x), t \geq 0)$. And from Proposition 8.8 we know that u_1 admits a variation of order $\frac{2\alpha}{\alpha+\gamma-d}$ which equal $c_{2,\alpha,\gamma}^{\frac{2\alpha}{\alpha+\gamma-d}}C_{\frac{1}{2},1-\frac{d-\gamma}{\alpha}}(A_2-A_1)$ with $C_{\frac{1}{2},1-\frac{d-\gamma}{\alpha}} = 2^{\frac{d-\gamma}{\alpha+\gamma-d}}\mu_{\frac{2\alpha}{\alpha+\gamma-d}}$ and means that

$$\sum_{i=0}^{n-1} |u_\theta(t_{j+1},x) - u_\theta(t_j,x)|^{\frac{2\alpha}{\alpha+\gamma-d}}$$

$$\to_{n\to\infty} c_{2,\alpha,\gamma}^{\frac{2\alpha}{\alpha+\gamma-d}} 2^{\frac{d-\gamma}{\alpha+\gamma-d}} \mu_{\frac{2\alpha}{\alpha+\gamma-d}}(\theta A_2 - \theta A_1)|\theta^{-\frac{1}{2}}|^{\frac{2\alpha}{\alpha+\gamma-d}}$$

$$= c_{2,\alpha,\gamma}^{\frac{2\alpha}{\alpha+\gamma-d}} 2^{\frac{d-\gamma}{\alpha+\gamma-d}} \mu_{\frac{2\alpha}{\alpha+\gamma-d}}(A_2 - A_1)|\theta|^{\frac{\gamma-d}{\alpha+\gamma-d}}.$$

$\square$

Recall that we show in Proposition 9.9 that for every fixed time $t > 0$,

$$\left(\theta^{\frac{1}{2}}\sqrt{C_{0,\alpha,\gamma}}^{-1}u_\theta(t,x), x \in \mathbb{R}^d\right)$$

is a perturbed multiparameter isotropic fractional Brownian motion as defined in Proposition 9.9. Then we can deduce the variation in space of u_θ, by recalling that $x_i = (x_i^{(1)}, \ldots, x_i^{(d)})$ with $x_i^{(j)} = A_1 + \frac{i}{n}(A_2 - A_1)$ for $i = 0, \ldots, n$ and $j = 1, \ldots, d$.

Proposition 9.10. *Let u_θ be given by (9.17). Then*

$$S_{[A_1,A_2]}^{N,\frac{2}{\alpha+\gamma-d}}(u(t,\cdot)) = \sum_{j=0}^{n-1} |u_\theta(t,x_{j+1}) - u_\theta(t,x_j)|^{\frac{2}{\alpha+\gamma-d}}$$

$$\to_{n\to\infty} \sqrt{C_{0,\alpha,\gamma}^{\frac{2}{\alpha+\gamma-d}}} d(A_2 - A_1)\mu_{\frac{2}{\alpha+\gamma-d}}|\theta|^{\frac{-1}{\alpha+\gamma-d}}$$

Proof. We use Proposition 9.9 and Lemma 8.6. $\square$

9.2.2   *Estimators of the drift parameter*

From relation (9.22) we can naturally define the following estimator for the parameter $\theta > 0$ of the stochastic partial differential equation (9.15)

$$\widehat{\theta}_{n,3} = \left(\left(c_{2,\alpha,\gamma}^{\frac{2\alpha}{\alpha+\gamma-d}} 2^{\frac{d-\gamma}{\alpha+\gamma-d}} \mu_{\frac{2\alpha}{\alpha+\gamma-d}}(A_2 - A_1)\right)^{-1} \right. \tag{9.23}$$

$$\left. \times \sum_{i=0}^{n-1} |u_\theta(t_{j+1},x) - u_\theta(t_j,x)|^{\frac{2\alpha}{\alpha+\gamma-d}}\right)^{\frac{\alpha+\gamma-d}{\gamma-d}}$$

$$= \left(c_{2,\alpha,\gamma}^{\frac{2\alpha}{\alpha+\gamma-d}} 2^{\frac{d-\gamma}{\alpha+\gamma-d}} \mu_{\frac{2\alpha}{\alpha+\gamma-d}}(A_2 - A_1)\right)^{\frac{d-\gamma}{\alpha+\gamma-d}}$$

$$\times \left(S^{n,\frac{2\alpha}{\alpha+\gamma-d}}(u_\theta(\cdot,x))\right)^{\frac{\alpha+\gamma-d}{\gamma-d}}$$

and so

$$\widehat{\theta}_{n,3}^{\frac{\gamma-d}{\alpha+\gamma-d}} = \frac{1}{c_{2,\alpha,\gamma}^{\frac{2\alpha}{\alpha+\gamma-d}} 2^{\frac{d-\gamma}{\alpha+\gamma-d}} \mu_{\frac{2\alpha}{\alpha+\gamma-d}} (A_2 - A_1)} \sum_{i=0}^{n-1} |u_\theta(t_{j+1}, x) - u_\theta(t_j, x)|^{\frac{2\alpha}{\alpha+\gamma-d}}.$$

(9.24)

We have the following asymptotic behavior.

Proposition 9.11. *Assume* $\frac{2\alpha}{\alpha+\gamma-d} := q$ *is an even integer and consider the estimator* $\widehat{\theta}_{n,3}$ *in (9.23). Then* $\widehat{\theta}_{n,3} \to_{n\infty} \theta$ *in probability and*

$$\sqrt{n} \left[\widehat{\theta}_{n,3}^{\frac{\gamma-d}{\alpha+\gamma-d}} - \theta^{\frac{\gamma-d}{\alpha+\gamma-d}} \right] \to^{(d)} N(0, s_{3,\theta,\alpha,\gamma}^2)$$

(9.25)

with

$$s_{3,\theta,\alpha,\gamma}^2 = \sigma_{\frac{1}{q},q}^2 \theta^{\frac{2(\gamma-d)}{\alpha+\gamma-d}} \mu_{\frac{2\alpha}{\alpha+\gamma-d}}^{-2}.$$

Proof. From Proposition 9.8 and the relation between the fractional and bifractional Brownian motion (see Theorem 2.2), we can see that

$$\left(c_{2,\alpha,\gamma}^{-1} 2^{\frac{d-\gamma}{2\alpha}} \theta^{\frac{d-\gamma}{2\alpha}} u_\theta(t,x), t \geq 0 \right)$$

is a perturbed fBm with Hurst parameter $H = \frac{\alpha-d+\gamma}{2\alpha}$. By taking $H = \frac{\alpha+\gamma-d}{2\alpha}$ and $q = \frac{1}{H} = \frac{2\alpha}{\alpha+\gamma-d}$ in Lemma 8.1, we get, as $n \to \infty$,

$$\frac{1}{\sqrt{n}} \sum_{i=0}^{n-1} \left[\frac{n\theta^{\frac{d-\gamma}{\alpha+\gamma-d}}}{c_{2,\alpha,\gamma}^{\frac{2\alpha}{\alpha+\gamma-d}} 2^{\frac{d-\gamma}{\alpha+\gamma-d}} (A_2 - A_1)} (u_\theta(t_{j+1}, x) - u_\theta(t_j, x))^{\frac{2\alpha}{\alpha+\gamma-d}} \right.$$

$$\left. -\mu_{\frac{2\alpha}{\alpha+\gamma-d}} \right] \to N(0, \sigma_{\frac{1}{q},q}^2)$$

or, equivalently

$$\sqrt{n}\mu_{\frac{2\alpha}{\alpha+\gamma-d}} \theta^{\frac{d-\gamma}{\alpha+\gamma-d}} \left[\widehat{\theta}_{n,3}^{\frac{\gamma-d}{\alpha+\gamma-d}} - \theta^{\frac{\gamma-d}{\alpha+\gamma-d}} \right]$$

$$\to_{n\to\infty} N(0, \sigma_{\frac{1}{q},q}^2)$$

which is equivalent to (9.10). $\qquad\square$

We finally obtain the asymptotic normality and the convergence in law for the estimator $\widehat{\theta}_{n,3}$.

Proposition 9.12. *Let* $\widehat{\theta}_{n,3}$ *be given by (9.23) and* $s_{3,\theta,\alpha,\gamma}$ *be given by (9.25). Then as* $n \to \infty$,

$$\sqrt{n} \left(\widehat{\theta}_{n,3} - \theta \right) \to^{(d)} N \left(0, s_{3,\theta,\alpha,\gamma} \left(\frac{\alpha+\gamma-d}{\gamma-d} \right)^2 \theta^{\frac{2\alpha}{\alpha+\gamma-d}} \right).$$

Proof. It suffices to apply (9.11) with $g(x) = x^{\frac{\alpha+\gamma-d}{\gamma-d}}$ and $\gamma_0 = \theta^{\frac{\gamma-d}{\gamma+\alpha-d}}$ and to follow the proof of Proposition 9.5. $\qquad\square$

We will repeat the method employed in the previous parts of our work in order to define an estimator expressed in terms of the variations in space of the process (9.17) for the parameter θ in (9.15).

Define, for every $n \geq 1$

$$\widehat{\theta}_{n,4} = \left[(\sqrt{C_{0,\alpha,\gamma}}^{\frac{2}{\alpha+\gamma-d}} \mu_{\frac{2}{\alpha+\gamma-d}} d(A_2 - A_1))^{-1} \right. \tag{9.26}$$

$$\left. \sum_{i=0}^{n-1} |u_\theta(t, x_{j+1}) - u_\theta(t, x_j)|^{\frac{2}{\alpha+\gamma-d}} \right]^{-(\alpha+\gamma-d)}$$

and so

$$\widehat{\theta}_{n,4}^{\frac{-1}{\alpha+\gamma-d}} = \frac{1}{\sqrt{C_{0,\alpha,\gamma}}^{\frac{2}{\alpha+\gamma-d}} \mu_{\frac{2}{\alpha+\gamma-d}} d(A_2 - A_1)} \sum_{i=0}^{n-1} |u_\theta(t, x_{j+1}) - u_\theta(t, x_j)|^{\frac{2}{\alpha+\gamma-d}}. \tag{9.27}$$

We can deduce the asymptotic properties of the estimator by using Proposition 9.9 and Lemma 8.6.

Proposition 9.13. *The estimator (9.26) converges in probability as* $n \to \infty$ *to the parameter* θ. *Moreover, if* $\frac{2}{\alpha+\gamma-d}$ *is an even integer, then*

$$\sqrt{n} \left[\widehat{\theta}_{n,4}^{\frac{-1}{\alpha+\gamma-d}} - \theta^{\frac{-1}{\alpha+\gamma-d}} \right] \to N(0, s_{4,\theta,\alpha,\gamma}^2)$$

with

$$s_{4,\theta,\alpha,\gamma}^2 = \sigma_{\frac{\alpha+\gamma-1}{2}, \frac{2}{\alpha+\gamma-d}}^2 \mu_{\frac{2}{\alpha+\gamma-d}}^{-2} \theta^{\frac{-2}{\alpha+\gamma-d}}.$$

Finally,

Proposition 9.14. *With* $\widehat{\theta}_{n,4}$ *from (9.26), as* $n \to \infty$,

$$\sqrt{n} \left(\widehat{\theta}_{n,4} - \theta \right) \to^{(d)} N \left(0, s_{4,\theta,\alpha,\gamma} \left(\frac{\alpha+\gamma-d}{\gamma-d} \right)^2 \theta^{\frac{2\alpha}{\alpha+\gamma-d}} \right).$$

Proof. Apply again (9.11) with $g(x) = x^{\frac{\alpha+\gamma-d}{\gamma-d}}$ and $\gamma_0 = \theta^{\frac{\gamma-d}{\gamma+\alpha-d}}$. $\qquad\square$

Remark 9.1. Notice that in the case $\gamma = 1$ (i.e. there is no spatial correlation and this case d has to be 1), we retrieve the results in Section 9.1.

Chapter 10

Power variations and inference for the stochastic wave equation

We come back to the wave equation described in Section 7. This time we will assume that the spatial dimension is $d = 1$. Recall the wave equation with space-time white noise, already defined in (7.1),

$$\begin{cases} \frac{\partial^2 u}{\partial t^2}(t,x) = \Delta u(t,x) + \dot{W}(t,x), & t > 0, \ x \in \mathbb{R} \\ u(0,x) = 0, & x \in \mathbb{R} \\ \frac{\partial u}{\partial t}(0,x) = 0, & x \in \mathbb{R}. \end{cases} \tag{10.1}$$

where W is a centered Gaussian field with covariance (3.5). Recall also that the mild solution $(u(t,x), t \geq 0, x \in \mathbb{R})$ to (10.1) is given by (7.4) and it is well-defined when the spatial dimension is one.

The purpose of this chapter is to analyze the asymptotic behavior of the p-variation for the solution to (10.1) and to the parametrized wave equation and to apply these results to the estimation of the drift and diffusion parameters of this equation.

10.1 Power variation of the solution

We will proceed as for the heat equation: we will analyze the variations of the solution in time and in space and we will apply the results obtained to parameter estimation.

10.1.1 *Temporal variation*

We first study the variation in time of the random field u, the mild solution to (10.1). Let $0 \leq A_1 < A_2$ be two real numbers. Following Definition 8.1, we define, for every $x \in \mathbb{R}$,

$$S_{[A_1,A_2]}^{N,2}(u(\cdot,x)) = \sum_{i=0}^{N-1} (u(t_{i+1},x) - u(t_i,x))^2 \qquad (10.2)$$

where $t_i = A_1 + \frac{i}{N}(A_2 - A_1), i = 0, 1, \ldots, N$ constitutes the partition (2.14) of the interval $[A_1, A_2]$. As in the rest of this monograph, the sequence $\left(S_{[A_1,A_2]}^{N,2}(u(\cdot,x)), N \geq 1\right)$ is called *the temporal quadratic of the random field u over the interval $[A_1, A_2]$*.

By relation (7.18) in Proposition 7.4 and by Proposition 7.5, we have the following result

Proposition 10.1. *Let $(u(t,x), t \geq 0, x \in \mathbb{R})$ be given by (7.4) and let $t_i, i = 0, \ldots, N$ be the partition of $[A_1, A_2]$ given by (2.14). Then for $i = 0, \ldots, N-1$ and for every $x \in \mathbb{R}$*

$$\mathbf{E}\left((u(t_{i+1},x) - u(t_i,x))(u(t_{j+1},x) - u(t_j,x))\right) = \begin{cases} \frac{1}{4}(t_{i+1}^2 - t_i^2), & \text{for } i = j \\ 0 & \text{for } i \neq j. \end{cases}$$
$$(10.3)$$

Proof. For $i = j$, we use the formula (7.18) with $t = t_{i+1}$ and $s = t_i$. The case $i \neq j$ is a direct consequence of Proposition 7.5. $\qquad\square$

The above formula is the key ingredient to derive the asymptotic behavior of the sequence (10.2).

Proposition 10.2. *Consider the random field u (7.4). For every $x \in \mathbb{R}$, the sequence $(S_{[A_1,A_2]}^{N,2}(u(\cdot,x)), N \geq 1)$ defined by (10.2) converges, as $N \to \infty$, to $\frac{1}{4}(A_2^2 - A_1^2)$ in $L^2(\Omega)$ and almost surely.*

Proof. Let us start with the $L^2(\Omega)$-convergence. First notice that by (10.3)

$$\mathbf{E}S_{[A_1,A_2]}^{N,2}(u(\cdot,x)) = \sum_{i=0}^{N-1} \mathbf{E}(u(t_{i+1},x) - u(t_i,x))^2$$

$$= \sum_{i=0}^{N-1} \frac{1}{4}(t_{i+1}^2 - t_i^2) = \frac{1}{2}(A_2^2 - A_1^2).$$

From Corollary 7.1, for every $i = 0, 1, \ldots, N-1$,

$$u(t_{i+1},x) - u(t_i,x) \sim N\left(0, \frac{1}{4}(t_{i+1}^2 - t_i^2)\right),$$

and so, if $Z \sim N(0, 1)$,

$$\mathbf{E} \left| S^{N,2}_{[A_1, A_2]}(u(\cdot, x)) - \frac{1}{4}(A_2^2 - A_1^2) \right|^2$$

$$= Var\left(S^{N,2}_{[A_1, A_2]}(u(\cdot, x)) \right) = Var\left(\sum_{i=0}^{N-1} (u(t_{i+1}, x) - u(t_i, x))^2 \right)$$

$$= \sum_{i=0}^{N-1} Var\left((u(t_{i+1}, x) - u(t_i, x))^2 \right) = \sum_{i=0}^{N-1} Var\left(\left(\sqrt{\frac{1}{4}(t_{i+1}^2 - t_i^2)} Z \right)^2 \right)$$

$$= Var(Z^2) \frac{1}{16} \sum_{i=0}^{N-1} \left(t_{i+1}^2 - t_i^2 \right)^2$$

$$\leq C \sum_{i=0}^{N-1} (t_{i+1} - t_i)^2 \leq C \frac{1}{N} \to_{N \to \infty} 0 \tag{10.4}$$

where we used the independence of the temporal increments of the solution u (see Proposition 7.5). This gives the convergence of $S^{N,2}_{[A_1, A_2]}(u(\cdot, x))$ to $\frac{1}{2}(A_2^2 - A_1^2)$ as $N \to \infty$. To get the almost sure convergence, we use a Borel-Cantelli argument. For every $\gamma > 0$ and $p \geq 1$ we have by hypercontractivity (this property can be used because the random variable $S^{N,2}_{[A_1, A_2]}(u(\cdot, x)))$ belongs to a finite sum of Wiener chaoses, see e.g. [Nourdin and Peccati (2012)])

$$P\left(\left| S^{N,2}_{[A_1, A_2]}(u(\cdot, x)) - \frac{1}{4}(A_2^2 - A_1) \right| \geq N^{-\gamma} \right)$$

$$\leq N^{\gamma p} \mathbf{E} \left| S^{N,2}_{[A_1, A_2]}(u(\cdot, x)) - \frac{1}{4}(A_2^2 - A_1^2) \right|^p$$

$$\leq C_p N^{\gamma p} \left(\mathbf{E} \left| S^{N,2}_{[A_1, A_2]}(u(\cdot, x)) - \frac{1}{4}(A_2^2 - A_1^2) \right|^2 \right)^{\frac{p}{2}} \leq C_p N^{p(\gamma - \frac{1}{2})},$$

the last bound being obtained by (10.4). By choosing p large enough et $\gamma \in (0, \frac{1}{2})$, we have

$$\sum_{N \geq 1} P\left(\left| S^{N,2}_{[A_1, A_2]}(u(\cdot, x)) - \frac{1}{4}(A_2^2 - A_1) \right| \geq N^{-\gamma} \right)$$

$$\leq C_p \sum_{N \geq 1} N^{p(\gamma - \frac{1}{2})} < \infty.$$

The conclusion follows by Borel-Cantelli lemma. $\qquad \square$

Let us prove a CLT for the temporal quadratic variation (10.2). Let us assume for simplicity $A_1 = 0, A_2 = 1$ (the general case follows similarly, but the proof is a bit longer due to the fact that the solution u does not have stationary increments in time). Let

$$S^{N,2}(u(\cdot, x)) := S_{[0,1]}^{N,2}(u(\cdot, x)) = \sum_{i=0}^{N-1} |u(t_{i+1}, x) - u(t_i, x)|^2$$

and set

$$U_N(u(\cdot, x)) = S^{N,2}(u(\cdot, x)) - \mathbf{E}S^{N,2}(u(\cdot, x)) = S^{N,2}(u(\cdot, x)) - \frac{1}{4}. \quad (10.5)$$

The above sequence satisfies the following limit theorem.

Theorem 10.1. *Let u be given by (7.4) and let $\left(U_{N,[A_1,A_2]}(u(\cdot, x)), N \geq 1\right)$ be given by (10.5). Then*

$$\sqrt{N}U_{N,x}(u(\cdot, x)) \to_{N \to \infty}^{(d)} N\left(0, \frac{1}{6}\right). \quad (10.6)$$

Proof. Due to the self-similarity in time of the solution (point 1. in Proposition 7.4), for every $x \in \mathbb{R}$ and $N \geq 1$,

$$U_N(u(\cdot, x)) =^{(d)} \frac{1}{N^2} \sum_{i=0}^{N-1} (u(i+1, x) - u(i, x))^2 - \frac{1}{4}$$

$$= \frac{1}{N^2} \sum_{i=0}^{N-1} \left[(u(i+1, x) - u(i, x))^2 - \frac{1}{4}((i+1)^2 - i^2) \right].$$

For $i = 0, 1, \ldots, N-1$, denote

$$z_i = (u(i+1, x) - u(i, x))^2 - \frac{1}{4}((i+1)^2 - i^2). \quad (10.7)$$

Then $(z_i, i = 0, 1, \ldots, N-1)$ are independent centered random variables, but not identically distributed. Let, for $i = 0, \ldots, N-1$,

$$\sigma_i^2 = Var(z_i) = \mathbf{E}\left(u(i+1, x) - u(i, x)\right)^4 - \left(\mathbf{E}\left(u(i+1, x) - u(i, x)\right)^2\right)^2.$$

Since $u(i+1, x) - u(i, x) \sim N(0, \frac{1}{4}(2i+1))$, we have

$$\sigma_i^2 = 2\left(\mathbf{E}\left(u(i+1, x) - u(i, x)\right)^2\right)^2 = \frac{1}{8}(2i+1)^2.$$

Let

$$s_N^2 := \sigma_0^2 + \ldots + \sigma_{N-1}^2 = \sum_{i=0}^{N-1} \frac{1}{8}(2i+1)^2$$

Then for N large s_N^2 behaves as $\frac{N^2}{6}$. Since clearly z_i has moments of all order and since

$$s_N^{-2} \sum_{k=0}^{N-1} \mathbf{E}|z_i|^4 \to_{N \to \infty} 0$$

the Lyapunov Central Limit Theorem (see e.g. [Bilingsley (1995)], page 362) implies that

$$\frac{\sum_{i=0}^{N-1} z_i}{s_N} \to_{N \to \infty} N(0,1)$$

and this gives easily the conclusion (10.6). $\qquad \square$

Remark 10.1. It is possible to prove a similar result on an arbitrary interval. Let $0 \leq A_1 < A_2$ and consider the partition (2.14). For $x \in \mathbb{R}$ and $N \geq 1$, let

$$U_{N,[A_1,A_2]}(u(\cdot,x)) = S_{[A_1,A_2]}^{N,2}(u(\cdot,x)) - \mathbf{E}S_{[A_1,A_2]}^{N,2}(u(\cdot,x))$$
$$= S_{[A_1,A_2]}^{N,2} - \frac{1}{4}(A_2^2 - A_1^2).$$

Then the sequence $\left(U_{N,[A_1,A_2]}(u(\cdot,x)), N \geq 1\right)$ satisfies

$$\sqrt{N}U_{N,[A_1,A_2]}(u(\cdot,x)) \to_{N \to \infty}^{(d)} N(0, K(A_1,A_2))$$

with

$$K(A_1, A_2) = \frac{A_2^3 - A_1^3}{6}(A_2 - A_1). \tag{10.8}$$

A proof of this limit theorem can be found in [Assaad *et al.* (2021)].

10.1.2 *Spatial variation*

Recall that the spatial covariance of the solution to the stochastic wave equation is given by (7.19). We will analyze the sequence

$$S_{[A_1,A_2]}^{N,2}(u(t,\cdot)) = \sum_{i=0}^{N-1} \left(u(t,x_{i+1}) - u(t,x_i)\right)^2 \tag{10.9}$$

with $t > 0$ fixed and $x_i, i = 0, \ldots, N$ given by (8.18).

From Proposition 7.6, we deduce the following result, which is crucial to get the asymptotic behavior of the sequence (10.9).

Lemma 10.1. *Let* $(u(t, x), t \geq 0, x \in \mathbb{R})$ *e be given by (7.4). Then, for* $t > \frac{1}{2}$, *we have for every* $i = 0, 1, \ldots, N - 1$,

$$\mathbf{E}\left[(u(t, x_{i+1}) - u(t, x_i))(u(t, x_{j+1}) - u(t, x_j))\right] \qquad (10.10)$$

$$= \begin{cases} \frac{t}{2N}(A_2 - A_1) - \frac{1}{8N^2}(A_2 - A_1)^2, & \text{if } i = j \\ -\frac{1}{8N^2}, & \text{if } i \neq j. \end{cases}$$

Proof. The case $i = j$ follows from (7.21). For $i \neq j$ (assume $i > j$), we have by (7.19),

$$\mathbf{E}\left[(u(t, x_{i+1}) - u(t, x_i))(u(t, x_{j+1}) - u(t, x_j))\right]$$

$$= \frac{1}{4}\left[2\left(\frac{i-j}{2N} - t\right)^2 - \left(\frac{i-j-1}{2N} - t\right)^2 - \left(\frac{i-j+1}{2N} - t\right)^2\right]$$

$$= -\frac{1}{8N^2}.$$

$$\square$$

By using the above lemma, we obtain the following limit theorem for the sequence (10.9).

Proposition 10.3. *Let* $\left(S^{N,2}_{[A_1, A_2]}(u(t, \cdot)), N \geq 1\right)$ *be given by (10.9) with* $t > 0$ *fixed. Then*

$$S^{N,2}_{[A_1, A_2]}(u(t, \cdot)) \to_{N \to \infty} \frac{t}{2}(A_2 - A_1) \text{ almost surely and in } L^2(\Omega).$$

Proof. By (10.1), we see that

$$\mathbf{E}S^{N,2}_{[A_1, A_2]}(u(t, \cdot)) = \frac{t}{2}(A_2 - A_1) - \frac{1}{8N}(A_2 - A_1)^2$$

and consequently

$$\mathbf{E}S^{N,2}_{[A_1, A_2]}(u(t, \cdot)) - \frac{t}{2}(A_2 - A_1) = -\frac{1}{8N}(A_2 - A_1)^2 \to_{N \to \infty} 0. \quad (10.11)$$

Now we write

$$S^{N,2}_{[A_1, A_2]}(u(t, \cdot)) - \frac{t}{2}(A_2 - A_1) = \left(S^{N,2}_{[A_1, A_2]}(u(t, \cdot)) - \mathbf{E}S^{N,2}_{[A_1, A_2]}(u(t, \cdot))\right)$$

$$+ \left(\mathbf{E}S^{N,2}_{[A_1, A_2]}(u(t, \cdot)) - \frac{t}{2}(A_2 - A_1)\right)$$

and then by (10.11)

$$\mathbf{E}\left(S^{N,2}_{[A_1,A_2]}(u(t,\cdot)) - \frac{t}{2}(A_2 - A_1)\right)^2$$

$$= \mathbf{E}\left(S^{N,2}_{[A_1,A_2]}(u(t,\cdot)) - \mathbf{E}S^{N,2}_{[A_1,A_2]}(u(t,\cdot))\right)^2$$

$$+ \left(\mathbf{E}S^{N,2}_{[A_1,A_2]}(u(t,\cdot)) - \frac{t}{2}(A_2 - A_1)\right)^2$$

$$= \mathbf{E}\left(S^{N,2}_{[A_1,A_2]}(u(t,\cdot)) - \mathbf{E}S^{N,2}_{[A_1,A_2]}(u(t,\cdot))\right)^2 + \frac{1}{64N^2}(A_2 - A_1)^4.$$

Using now (10.10),

$$\mathbf{E}\left(S^{N,2}_{[A_1,A_2]}(u(t,\cdot)) - \mathbf{E}S^{N,2}_{[A_1,A_2]}(u(t,\cdot))\right)^2$$

$$= 2\sum_{i=0}^{N-1}\left(\frac{t}{2N}(A_2 - A_1) - \frac{1}{8N^2}(A_2 - A_1)^2\right)^2 + 2\sum_{i,j=0,i\neq j}^{N-1}\left(-\frac{1}{8N^2}\right)^2$$

$$= \frac{t^2}{2N}(A_2 - A_1)^2 + o\left(\frac{1}{N}\right)$$

and thus,

$$\mathbf{E}\left(S^{N,2}_{[A_1,A_2]}(u(t,\cdot)) - \frac{t}{2}(A_2 - A_1)\right)^2 \tag{10.12}$$

$$= \frac{t^2}{2N}(A_2 - A_1)^2 + o\left(\frac{1}{N}\right) \to_{N\to\infty} 0.$$

The almost sure convergence follows as in the proof of Proposition 10.2, via Borel-Cantelli and hypercontractivity. Indeed, for $0 < \gamma < \frac{1}{2}, p \geq 2$

$$P\left(\left|S^{N,2}_{[A_1,A_2]}(u(t,\cdot)) - \frac{t}{2}(A_2 - A_1)\right| \geq N^{-\gamma}\right)$$

$$\leq N^{\gamma p}\mathbf{E}\left|S^{N,2}_{[A_1,A_2]}(u(t,\cdot)) - \frac{t}{2}(A_2 - A_1)\right|^p$$

$$\leq C_p N^{\gamma p}\left[\mathbf{E}\left|S^{N,2}_{[A_1,A_2]}(u(t,\cdot)) - \frac{t}{2}(A_2 - A_1)\right|^2\right]^{\frac{p}{2}} \leq C_p N^{(\gamma-\frac{1}{2})p}$$

where we used (10.12) for the last inequality. Thus, since $\gamma < \frac{1}{2}$, by taking p large enough,

$$\sum_{N\geq 1} P\left(\left|S^{N,2}_{[A_1,A_2]}(u(t,\cdot)) - \frac{t}{2}(A_2 - A_1)\right| \geq N^{-\gamma}\right)$$

$$\leq C_p \sum_{N\geq 1} N^{(\gamma-\frac{1}{2})p} < \infty.$$

$\square$

It is also possible to prove that the sequence (10.9) satisfies a CLT. We refer to [Khalil *et al.* (2018)] for the proof (see also [Assaad *et al.* (2021)]).

Theorem 10.2. *Let* $\left(S^{N,2}_{[A_1,A_2]}(u(t,\cdot)), N \geq 1\right)$ *be given by (10.9). Then*

$$\sqrt{N}\left(S^{N,2}_{[A_1,A_2]}(u(t,\cdot)) - \frac{t}{2}(A_2 - A_1)\right) \to^{(d)}_{N\to\infty} N(0, \sigma^2(A_1, A_2)$$

with $\sigma^2(A_1, A_2) = \frac{t^2}{2}(A_2 - A_1)^2 > 0.$

10.2 The parametrized wave equation with space-time white noise

Let us now consider the parametrized linear stochastic wave equation

$$\frac{\partial^2 u_\theta}{\partial t^2} = \theta \Delta u_\theta(t, x) + \dot{W}(t, x), \qquad t \geq 0, x \in \mathbb{R} \qquad (10.13)$$

with vanishing initial condition

$$u_\theta(0, x) = \frac{\partial u_\theta}{\partial t}(0, x) = 0 \text{ for every } x \in \mathbb{R}.$$

We assume $\theta > 0$ and W is the space-time white noise defined by its covariance (3.5). The purpose is to estimate the parameter θ on the basis of the observations $(u(t_i, x), i = 0, \ldots, N)$ at a fixed point in space $x \in \mathbb{R}$. We assume that the observation times $t_i, i = 0, 1, \ldots, N$ are given by (2.14). Another possibility is to assume that we observe $(u(t, x_i), i = 0, 1, \ldots, N)$ with $t > 0$ fixed and $x_i, i = 0, 1, \ldots, N$ given by (8.18).

We start with a brief study of the solution to (10.13), then we define an estimator the drift parameter $\theta > 0$ in the SPDE (10.13).

10.2.1 *Properties of the solution*

The Green kernel associated to the SPDE (10.13), i.e. the deterministic function which solves $\frac{\partial^2}{\partial t^2} v(t, x) = \theta \Delta v(t, x)$ is $G_1(\sqrt{\theta}t, x)$ with G_1 defined by (7.2). Therefore, the mild solution to the SPDE (10.13) can be written as

$$u_\theta(t, x) = \int_0^t \int_{\mathbb{R}} G_1(\sqrt{\theta}(t - s, x - y))W(ds, dy) \qquad (10.14)$$

for every $t \geq 0, x \in \mathbb{R}$. For $d = 1$, the above Wiener integral with respect to the Gaussian noise W is well-defined (Proposition 7.1).

We consider the following transform of the mild solution (10.14). Denote by

$$v_\theta(t, x) = u_\theta\left(\frac{t}{\sqrt{\theta}}, x\right), \qquad (t, x) \in [0, \infty) \times \mathbb{R}. \qquad (10.15)$$

We notice the following property for the random field v_θ.

Lemma 10.2. *The random field $(v_\theta(t, x), t \geq 0, x \in \mathbb{R})$ is a mild solution to the stochastic heat equation*

$$\frac{\partial}{\partial t} v_\theta(t, x) = \Delta v_\theta(t, x) + \theta^{-\frac{1}{4}} \widetilde{W}(t, x)$$

with $\widetilde{W}$ a space-time white noise, i.e. for every $t \geq 0, x \in \mathbb{R}$,

$$v_\theta(t, x) = \theta^{-\frac{1}{4}} \int_0^t \int_{\mathbb{R}} G_1(t - s, x - y) \widetilde{W}(ds, dy). \qquad (10.16)$$

Proof. We have

$$\begin{aligned}
v_\theta(t, x) &= \int_0^{\frac{t}{\sqrt{\theta}}} \int_{\mathbb{R}} G_1(t - \sqrt{\theta}s, x - y) W(ds, dy) \\
&= \int_0^t \int_{\mathbb{R}} G_1(t - s, x - y) W\left(d\left(\frac{s}{\sqrt{\theta}}\right), dy\right) \\
&= \theta^{-\frac{1}{4}} \int_0^t \int_{\mathbb{R}} G_1(t - s, x - y) \widetilde{W}(ds, dy)
\end{aligned}$$

where $(\widetilde{W}(t, x), t \geq 0, x \in \mathbb{R})$ with $\widetilde{W}(t, x) = \theta^{\frac{1}{4}} W\left(\frac{t}{\sqrt{\theta}}, x\right)$ is a space-time white noise. $\qquad \square$

Let us give the asymptotic behavior of the temporal quadratic variation of the random field u_θ.

Proposition 10.4. *Let $(t_i, i = 0, 1, \ldots, N)$ be the partition of $[A_1, A_2]$ given by (2.14) and let u_θ be defined by (10.14). Denote, for $N \geq 1$, $x \in \mathbb{R}$,*

$$S_{[A_1, A_2]}^{N,2}(u_\theta(\cdot, x)) = \sum_{i=0}^{N-1} (u_\theta(t_{i+1}, x) - u_\theta(t_i, x))^2. \qquad (10.17)$$

Then

$$S_{[A_1, A_2]}^{N,2}(u_\theta(\cdot, x)) \to_{N \to \infty} \frac{1}{4}(A_2^2 - A_1^2)\theta^{\frac{1}{2}} \text{ almost surely and in } L^2(\Omega).$$

Proof. Notice that, by (10.16),

$$\left(\theta^{\frac{1}{4}} v_\theta(t, x), t \geq 0, x \in \mathbb{R}\right) \equiv^{(d)} \left(u_1(t, x), t \geq 0, x \in \mathbb{R}\right) \qquad (10.18)$$

where $\equiv^{(d)}$ means the equivalence of finite dimensional distributions (while $=^{(d)}$ stands for the equality in law). Then, by the definition of v_θ (see (10.15)), (10.18) and Corollary 1.5,

$$
\begin{aligned}
S^{N,2}_{[A_1, A_2]}(u_\theta(\cdot, x)) &= \sum_{i=0}^{N-1} \left(v_\theta(\sqrt{\theta} t_{i+1}, x) - v_\theta(\sqrt{\theta} t_i, x)\right)^2 \\
&= \theta^{-\frac{1}{2}} \sum_{i=0}^{N-1} \left(\theta^{\frac{1}{4}} v_\theta(\sqrt{\theta} t_{i+1}, x) - \theta^{\frac{1}{4}} v_\theta(\sqrt{\theta} t_i, x)\right)^2 \\
&=^{(d)} \theta^{-\frac{1}{2}} \sum_{i=0}^{N-1} \left(\tilde{u}_1(\sqrt{\theta} t_{i+1}, x) - \tilde{u}_1(\sqrt{\theta} t_i, x)\right)^2
\end{aligned}
$$

with

$$\tilde{u}_1(t, x) = \int_0^t \int_{\mathbb{R}} G_1(t - s, x - y) \widetilde{W}(ds, dy), \qquad (10.19)$$

where $\widetilde{W}$ is the space-time white noise from (10.16).

Since $(\sqrt{\theta} t_i, i = 0, \ldots, N)$ represents a partition of the interval $[\sqrt{\theta} A_1, \sqrt{\theta} A_2]$, we have by Proposition 10.3 that

$$
\begin{aligned}
\sum_{i=0}^{N-1} &\left(\tilde{u}_1(\sqrt{\theta} t_{i+1}, x) - \tilde{u}_1(\sqrt{\theta} t_i, x)\right)^2 \\
&\to_{N \to \infty} \frac{1}{4} \left((\sqrt{\theta} A_2)^2 - (\sqrt{\theta} A_1)^2\right) = \frac{1}{4} \theta(A_2^2 - A_1^2).
\end{aligned}
$$

This implies the conclusion. $\qquad \square$

Remark 10.2. Notice that, for $x \in \mathbb{R}$ and $\theta > 0$, $(u_\theta(t, x), t \geq 0)$ is a Gaussian process with covariance

$$
\begin{aligned}
\mathbf{E} u_\theta(t, x) u_\theta(s, x) &= \mathbf{E} v_\theta(\sqrt{\theta} t, x) v_\theta(\sqrt{\theta} s, x) \\
&= \theta^{-\frac{1}{2}} \mathbf{E} u_1(\sqrt{\theta} t, x) u_1(\sqrt{\theta} s, x) = \theta^{-\frac{1}{2}} \frac{1}{4} \left((\sqrt{\theta} t) \wedge (\sqrt{\theta} s)\right)^2 \\
&= \frac{1}{4} \theta^{\frac{1}{2}} (t \wedge s)^2.
\end{aligned}
$$

Actually, the process $(u_\theta(t, x), t \geq 0)$ coincides in distribution with

$$\left(B_{\frac{1}{2} \theta^{\frac{1}{4}} t^2}, t \geq 0\right)$$

where B is a standard Brownian motion.

Let us now discuss the quadratic variation in space of the solution to the SPDE (10.13). We fix $t > 0$ and we analyze the process $(u_\theta(t, x), x \in \mathbb{R})$. By Lemma 10.2 and (7.19), its covariance can be computed as follows, for $t > 0, x, y \in \mathbb{R}$,

$$\mathbf{E}u_\theta(t, x)u_\theta(t, y) = \mathbf{E}v_\theta(\sqrt{\theta}t, x)v_\theta(\sqrt{\theta}t, y)$$
$$= \theta^{-\frac{1}{2}}\mathbf{E}u_1(\sqrt{\theta}t, x)u_1(\sqrt{\theta}t, y)$$

so for every $t \geq 0, x, y \in \mathbb{R}$

$$\mathbf{E}u_\theta(t, x)u_\theta(t, y) = \frac{1}{4\sqrt{\theta}}\left(\frac{|x-y|}{2} - t\sqrt{\theta}\right)^2 1_{|x-y| \leq t\sqrt{\theta}}. \qquad (10.20)$$

Define for every $N \geq 1$ and $x \in \mathbb{R}$,

$$S_{[A_1, A_2]}^{N,2}(u_\theta(t, \cdot)) = \sum_{i=0}^{N-1}(u_\theta(t, x_{i+1}) - u_\theta(t, x_i))^2 \qquad (10.21)$$

with $(x_i, i = 0, \ldots, N)$ given by (8.18). If $2t\sqrt{\theta} > 1$, we find from (10.20),

$$\mathbf{E}(u_\theta(t, x_{i+1}) - u_\theta(t, x_i))^2 = \frac{t}{2N}(A_2 - A_1) - \frac{1}{8\sqrt{\theta}N^2}(A_2 - A_1)^2.$$

Therefore,

$$\mathbf{E}S_{[A_1, A_2]}^{N,2}(u_\theta(t, \cdot)) = \frac{t}{2}(A_2 - A_1) - \frac{1}{8\sqrt{\theta}N}(A_2 - A_1)^2.$$

Proposition 10.5. *Consider the sequence* $\left(S_{[A_1, A_2]}^{N,2}(u_\theta(t, \cdot)), N \geq 1\right)$ *given by (10.21). Then*

$$S_{[A_1, A_2]}^{N,2}(u_\theta(t, \cdot)) \to_{N \to \infty} \frac{t}{2}(A_2 - A_1) \text{ almost surely and in } L^2(\Omega). \quad (10.22)$$

Proof. It suffices to follow the lines of the proof of Proposition 10.4. $\qquad \square$

10.2.2 Estimation of the drift parameter

By Proposition 10.4 we define the following estimator for the parameter $\theta > 0$ in (5.16)

$$\widehat{\theta}_N := \left(\frac{4}{A_2^2 - A_1^2}S_{[A_1, A_2]}^{N,2}(u_\theta(\cdot, x))\right)^2 \qquad (10.23)$$

with $S_{[A_1,A_2]}^{N,2}(u_\theta(\cdot,x))$ given by (10.17). Notice that the estimator $\widehat{\theta}_N$ can be computed from the discrete observations $u_\theta(t_i,x)$ with t_i defined by (2.14).

The following result is immediate.

Proposition 10.6. *The estimator $\widehat{\theta}_N$ is strongly consistent, that is, $\widehat{\theta}_N \to_{N\to\infty} \theta$ almost surely.*

Proof. It follows directly from Proposition 10.4. $\qquad\qquad\square$

The next step is to show that the estimator $\widehat{\theta}_N$ is asymptotically Gaussian.

Theorem 10.3. *Let $0 \le A_1 < A_2$ and let $K(A_1, A_2)$ be given by (10.8). Consider the estimator $\widehat{\theta}_N$ given by (10.23). Then*

$$\sqrt{N}\left(\widehat{\theta}_N^{\frac{1}{2}} - \theta^{\frac{1}{2}}\right) \to_{N\to\infty}^{(d)} N(0, H(\theta, A_1, A_2)) \qquad (10.24)$$

where $H(\theta, A_1, A_2) = \left(\frac{4}{A_2^2 - A_1^2}\theta^{-\frac{1}{2}}\right)^2 K(\sqrt{\theta}A_1, \sqrt{\theta}A_2)$.

Proof. We write, for every $N \ge 1$, with $\tilde{u}_1$ from (10.19),

$$\widehat{\theta}_N^{\frac{1}{2}} - \theta^{\frac{1}{2}}$$

$$= \frac{4}{A_2^2 - A_1^2} \sum_{i=0}^{N-1} \left(u_\theta(t_{i+1},x) - u_\theta(t_i,x)\right)^2 - \theta^{\frac{1}{2}}$$

$$= \frac{4}{A_2^2 - A_1^2} \sum_{i=0}^{N-1} \left(\tilde{u}_1(\sqrt{\theta}t_{i+1},x) - \tilde{u}_1(\sqrt{\theta}t_i,x)\right)^2 - \theta^{\frac{1}{2}}$$

$$= \frac{4}{A_2^2 - A_1^2}\theta^{-\frac{1}{2}}\left[\sum_{i=0}^{N-1}\left(\tilde{u}_1(\sqrt{\theta}t_{i+1},x) - \tilde{u}_1(\sqrt{\theta}t_i,x)\right)^2 - \frac{1}{4}(A_2^2 - A_1^2)\theta\right].$$

Now by Theorem 10.1, since $(\sqrt{\theta}t_i, i = 0, 1, \ldots, N)$ forms a partition of the interval $[\sqrt{\theta}A_1, \sqrt{\theta}A_2]$, we get

$$\sqrt{N}\left[\sum_{i=0}^{N-1}\left(\tilde{u}_1(\sqrt{\theta}t_{i+1},x) - \tilde{u}_1(\sqrt{\theta}t_i,x)\right)^2 - \frac{1}{4}(A_2^2 - A_1^2)\theta\right]$$
$$\to_{N\to\infty}^{(d)} N(0, K(\sqrt{\theta}A_1, \sqrt{\theta}A_2))$$

and this gives the conclusion. $\qquad\qquad\square$

By applying the delta-method, we can obtain a result similar to the limit theorem (8.17) with $\widehat{\theta}_N$ instead of $\widehat{\theta}_N^{\frac{1}{2}}$.

Corollary 10.1. *Let $\widehat{\theta}_N$ be given by (10.23). Then*

$$\sqrt{N}(\widehat{\theta}_N - \theta) \to N(0, L(\theta, A_1, A_2)) \tag{10.25}$$

with $L(\theta, A_1, A_2) = 4\theta H(\theta, A_1, A_2)$ with H given by (10.24).

Proof. Consider the function $g : \mathbb{R} \to \mathbb{R}$, $g(x) = x^2$. We can write, for every $N \geq 1$,

$$\sqrt{N}(\widehat{\theta}_N - \theta) = \sqrt{N}\left(g(\widehat{\theta}_N^{\frac{1}{2}}) - g(\theta^{\frac{1}{2}})\right).$$

Since $g'(\sqrt{\theta})$ does not vanish, it follows by the delta-method (see (9.11)) that

$$\sqrt{N}\left(g(\widehat{\theta}_N^{\frac{1}{2}}) - g(\theta^{\frac{1}{2}})\right) \xrightarrow[N\to\infty]{(d)} N(0, H(A_1, A_2)g'(\sqrt{\theta})^2)$$
$$= N(0, L(\theta, A_1, A_2)).$$

$\square$

Let us finish with some comments concerning the parameter estimator via the spatial quadratic variations. Recall the result stated in (10.22)

$$S_{[A_1,A_2]}^{N,2}(u_\theta(t, \cdot)) \to_{N\to\infty} \frac{t}{2}(A_2 - A_1) \text{ almost surely and in } L^2(\Omega)$$

where $S_{[A_1,A_2]}^{N,2}(u_\theta(t, \cdot))$ is given by (10.21).

From (10.22), we notice that the spatial quadratic variation $S_{N,t}(u_\theta)$ does not depend asymptotically on θ and it cannot be used to estimate the drift parameter in (10.13). On the other hand, it can still be used to estimate the diffusion parameter for the wave equation. Indeed, let us consider the SPDE

$$\frac{\partial^2 u_\lambda}{\partial t^2} = \Delta u_\lambda(t, x) + \lambda \dot{W}(t, x), \quad t \geq 0, x \in \mathbb{R} \tag{10.26}$$

with vanishing initial condition

$$u_\lambda(0, x) = \frac{\partial u_\lambda}{\partial t}(0, x) = 0 \text{ for every } x \in \mathbb{R}.$$

We assume again $\lambda > 0$ and W is the space-time white. The mild solution to (10.26) is given, for $t \geq 0, x \in \mathbb{R}$, by

$$u_\lambda(t, x) = \lambda \int_0^t \int_{\mathbb{R}} G_1(t - s, x - y)W(ds, dy) = \lambda u(t, x)$$

with G_1 from (7.2) and u from (7.4). It is well-defined since we consider the spatial dimension $d = 1$. Moreover,

$$S^{N,2}(u_\lambda(t,\cdot)) := \sum_{i=0}^{N-1} [u_\lambda(t, x_{i+1}) - u_\lambda(t, x_i)]^2 = \lambda^2 S^{N,2}(u(t,\cdot))$$

with $S^{N,2}(u(t,\cdot))$ given by (10.21) with $A_1 = 0, A_2 = 1$. Therefore, by Proposition 10.3,

$$S^{N,2}(u_\lambda(t,\cdot)) \to_{N \to \infty} \lambda^2 \frac{t}{2}(A_2 - A_1) \text{ almost surely and in } L^2(\Omega).$$

Consequently, as $N \to \infty$,

$$\widehat{\lambda}_N^2 := \frac{2S^{N,2}(u_\lambda(t,\cdot))}{t} \to \lambda^2 \text{ almost surely}$$

so $\widehat{\lambda}_N$ constitutes a consistent estimator for the diffusion parameter λ in (10.26). A CLT for this estimator can be deduced via Theorem 10.2.

Bibliography

B. J. Adler (1990). An introduction to Continuity, Extrema and Related Topics for General Gaussian processes. *Institute of Mathematical Statistics Lecture Notes-Monographs Series* **12**.

S. Asmussen (2005). Some Applications of the Delta Method. Lecture notes. Aarhus University.

O. Assaad, J. Gamain and C. A. Tudor (2021). Drift parameter estimation for the stochastic wave equation. Preprint.

A. Ayache, S. Leger and M. Pontier (2002). Drap Brownien fractionnaire, *Potential Analysis*, **17**(1), 31-43.

R. M. Balan and C. A. Tudor (2008). The stochastic heat equation with fractional-colored noise: existence of the solution, *Latin Amer. J. Probab. Math. Stat.* **4**, 57-87.

R. M. Balan and C. A. Tudor (2010). The stochastic wave equation with fractional noise: A random field approach, *Stoch. Proc. Appl.* **120**, 2468-2494.

P. Bilingsley (1995). Probability and measure. Wiley series in Probability and Mathematical Statistics. *IWiley.*

P. Breuer and P. Major (1983). Central limit theorems for nonlinear functionals of Gaussian fields, *J. Multivariate Analysis* **13**, 425-441.

E. Cabana (1970). The vibrating string forced by white noise, *Z. Wahrscheinlichkeits theorie Verw Geb.* **15**, 111-130.

P-L. Chow (2014). *Stochatic partial differential equations. Second Edition* (Taylor and Francis).

G. Da Prato and J. Zabczyk (1992). *Stochastic Equations in Infinite Dimensions* (Cambridge University Press, Cambridge).

J. Clarke De la Cerda, C. A. Tudor (2014). Hitting probabilities for the stochastic wave equation with fractional colored noise, *Rv. Mat. Iberoam.* **30**(2), 685-709.

L. Chen and R. Dalang (2015). Moments, intermitemcy and growth indices for nonlinear stochastic fractional heat equation, *Stoch. Partial Differ. Equ. Anal. Comput.*, **3**(3), 360-397.

R. C. Dalang (1999). Extending the martingale measure stochastic integral with applications to spatially homogeneous SPDE's, *Electr. J. Probab.* **4**,

1-29 pp. Erratum in *Electr. J. Probab.* **6** (2001), 5 pp.

R. C. Dalang, M. Sanz-Solé (2009). Hölder-Sobolev regularity of the solution to the stochastic wave equation in dimension three, *Memoirs of the American Mathematical Society*, **199**, 931, 1-70.

L. Debbi and M. Dozzi (2005). On the solutions of nonlinear stochastic fractional partial differential equations in one spatial dimension, *Stoch. Proc. Appl.*, **115**, 1761-1781.

R. L. Dobrushin and P. Major (1979). Non-central limit theorems for non-linear functionals of Gaussian fields, *Z. Wahrscheinlichkeitstheorie verw. Gebiete*, **50**, 27-52.

R. M. Dudley (2003). *Real Analysis and Probability. Second Edition* (Cambridge University Press, Cambridge).

P. Embrechts and M. Maejima (2002). *Selfsimilar processes* (Princeton University Press, Princeton, New York).

L. C. Evans (2010). Partial differential equations. *Graduate Studies in Mathematics*, **19**, AMS.

G. B. Folland (1975). *Introduction to Partial Differential Equations* (Princeton Univ. Press, Princeton).

M. Foondun, P. Mahboubi and D. Khoshnevisan (2015). Analysis of the gradient of the solution to a stochastic heat equation via fractional Brownian motion, *Stoch. Partial Differ. Equ. Anal. Comput.*, **3**(2), 133-158.

Giraitis, L. and Surgailis, D. (1990). A central limit theorem for quadratic forms in strongly dependent linear variables and its applications to the asymptotic normality of Whittle estimate, *Prob. Th. and Rel. Field.* **86**, 87-104.

D. Harnett and D. Nualart (2017). Decomposition and Limit Theorems for a Class of Self-Similar Gaussian Processes. *Stochastic Analysis and Related Topics: A Festschrift in Honor of Rodrigo Bañuelos, Springer International Publishing, Cham.*

E. Herbin (2006). From N-parameter fractional Brownian motion to N-parameter multifractional Brownian motion, *Rochy Mountain Journal of Mathematics*, **36**(4), 1249-1284.

N. Jacob and H. G. Leopold (1993). Pseudo differential operators with variable order of differentiation generating Feller semigroups, *Integral Equations Operator Theory*, **17**, 544-553.

N. Jacob, A. Potrykus and J.-L. Wu (2010). Solving a non-linear stochastic pseudo-differential equation of Burgers type, *Stochastic Process. Appl.*, **120**, 2447-2467.

Y. Jiang, K. Shi and Y. Wang (2010). Stochastic fractional Anderson models with fractional noises, *Chin. Ann. Math.*, **31B**(1), 101-118.

C. Jost (2006). Transformation formulas for fractional Brownian motion, *Stochastic Process. Appl.*, **116**, 1341-1357.

J. P. Kahane (1985). *Some Random Series of Functions* (Cambridge University Press, Cambridge).

I. Karatzas and S. Shreve (1988). *Brownian motion and stochastic calculus.* Graduate Texts in Mathematics, **113**. Springer-Verlag, New York, 1988.

M. Khalil and C. A. Tudor (2019). Correlation structure, quadratic variations and parameter estimation for the solution to the wave equation with fractional noise, *Electron. J. Stat.* **12**(2), 3639-3672.

M. Khalil, C. A. Tudor and M. Zili (2018). Spatial variation for the solution to the stochastic linear wave equation driven by additive space-time white noise, *Stoch. Dyn.* **18**(5), 1850036, 20 pp.

D. Khosnevisan (2002). *Multiparemeter processes. An introduction to random fields* (Springer, New York).

A. N. Kolmogorov (1950). *Foundations of Probability Theory* (Chelsea Publishing Co., New York).

P. Lei and D. Nualart (2009). A decomposition of the bifractional Brownian motion and some applications, *Statist. Probab. Lett.*, **79**(5), 619-624.

J. Liu and C. A. Tudor (2016). Central limit theorem for the solution to the heat equation with moving time, *Infin. Dimens. Anal. Quantum Probab. Relat. Topics.* **19**(1).

W. Liu and M. Röckner (2015). *Stochastic Partial Differential Equations: An Introduction* (Springer).

Z. Mahdi Khalil and C. A. Tudor (2019). On the distribution and q-variation of the solution to the heat equation with fractional Laplacian, *Probab. Math. Statist.*, **39**(2), 315-335.

Z. Mahdi Khalil and C. A. Tudor (2019). Estimation of the drift parameter for the fractional stochastic heat equation via power variation, *Mod. Stoch. Theory Appl.* **6**(4), 397-417.

V. Makogin and Y. Mishura (2019). Gaussian multi-self-similar random fields with distinct stationary properties of their rectangular increments, *Stoch. Models*, **35**(4), 391-428.

Y. Mishura (2008). *Stochastic Calculus for Fractional Brownian Motion and Related Processes* (Springer).

C. Mueller and R. Tribe (2002). Hitting probabilities of a random string, *Electronic J. Probab.*, **7**, Paper No. 10, 29 pp.

C. Mueller and Z. Wu (2009). A connection between the stochastic heat equation and fractional Brownian motion, and a simple proof of a result of Talagrand, *Electronic Comm. Probab.*, **14**(6), 55-65.

I. Nourdin (2012). *Selected topics on Fractional Brownian motion* (Springer and Bocconi).

I. Nourdin and G. Peccati (2012). *Normal Approximations with Malliavin Calculus From Stein's Method to Universality* (Cambridge University Press, Cambridge).

I. Nourdin, D. Nualart and C. A. Tudor (2010). Central and non-central limit theorems for weighted power variations of fractional Brownian motion, *Annales de l' Institut H. Poincaré*, **46**(4), 1055-1079.

D. Nualart (2006). *Malliavin Calculus and Related Topics. Second Edition* (Springer New York).

D. Nualart and S. Ortiz-Latorre (2008). Central limit theorems for multiple stochastic integrals multiple stochastic integral and Malliavin calculus, *Stochastic Processes and their Applications*, **118**, 614-628.

D. Nualart and G. Peccati (2005). Central limit theorems for sequences of multiple stochastic integrals multiple stochastic integral, *The Annals of Probability*, **33**(1), 173-193.

G. Peccati and M. S. Taqqu (2010). *Wiener Chaos: Moments, cumulants and Diagrams* (Springer Verlag, Bocconi and Springer Series).

G. Peccati and C. A. Tudor (2004). Gaussian limits for vector-valued multiple stochastic integrals, *Séminaire de Probabilités*, XXXIV, 247-262.

V. Pipiras and Murad Taqqu (2001). Integration questions related to the fractional Brownian motion, *Probability Theory and Related Fields*, **118**(2), 251-281.

J. Pospisil and R. Tribe (2007). Parameter estimates and exact variations for stochastic heat equations driven by space-time white noise, *Stoch. Anal. Appl.*, **25**(3), 593-611.

B. Oksendal (2010). *Stochastic partial differential equation. An introduction with applications* (Springer, Heidelberg New York).

D. Revuz and M. Yor (1994). *Continuous martingales and Brownian motion* (Springer-Verlag, Berlin).

F. Russo and C. A. Tudor (2006). On the bifractional Brownian motion, *Stochastic Process. Appl.*, **5**, 830-856.

G. Samorodnitsky and M. Taqqu (1994). *Stable Non-Gaussian random variables* (Chapman and Hall, London).

G. Samorodnitsky (2016). *Stochastic processes and long range dependence* (Springer Series in Operations Research and Financial Engineering. Springer, Cham.)

Swanson J. (2007). Variations of the solution to a stochastic heat equation, *Annals of Probability*, **35**(6), 2122-2159.

M. Taqqu (1979). Convergence of integrated processes of arbitrary Hermite rank, *Z. Wahrscheinlichkeitstheorie verw. Gebiete*, **50**, 53-83.

Treves, F. (1975). *Basic Linear Partial Differential Equations* (Academic Press, New York).

C. A. Tudor (2013). *Analysis of variations for self-similar processes. A stochastic calculus approach* (Probability and its Applications, Springer, Cham.)

C. A. Tudor and Y. Xiao (2017). Sample paths of the solution to the fractional-colored stochastic heat equation, *Stoch. Dyn.*, **17**(1), 1750004, 20 pp.

M. Zili and E. Zougar (2019). Exact variations for stochastic heat equations with piecewise constant coefficients and applications to parameter estimation, *Teor. Imovīr. Mat. Stat.* **100**, 75-101.

Printed in the United States
by Baker & Taylor Publisher Services